Compactifications, Configurations, and Cohomology

CONTEMPORARY MATHEMATICS

790

Compactifications, Configurations, and Cohomology

Conference
Compactifications, Configurations, and Cohomology
October 22–24, 2021
Northeastern University, Boston, Massachusetts

Peter Crooks
Alexandru I. Suciu
Editors

AMERICAN MATHEMATICAL SOCIETY

2020 *Mathematics Subject Classification.* Primary 14J42, 14L17, 14L30, 14M15, 14N20, 17B63, 32S22, 55N10, 55R80, 55U05.

Library of Congress Cataloging-in-Publication Data

Names: Conference on compactifications, configurations, and cohomology (2021 : Boston, Mass.), author | Crooks, Peter, 1987– editor. | Suciu, Alexandru I., editor.

Title: Compactifications, configurations, and cohomology / Peter Crooks, Alexandru I. Suciu, editors.

Description: Providence, Rhode Island : American Mathematical Society, 2023. | Series: Contemporary mathematics, 0271-4132 ; volume 790 | Includes bibliographical references.

Identifiers: LCCN 2023014589 | ISBN 9781470469924 (paperback) | ISBN 9781470474577 (ebook)

Subjects: LCSH: Geometry, Algebraic–Congresses. | Algebraic topology–Congresses. | AMS: Algebraic geometry – Algebraic groups – Affine algebraic groups, hyperalgebra constructions. | Algebraic geometry – Algebraic groups – Group actions on varieties or schemes (quotients). | Algebraic geometry – Special varieties – Grassmannians, Schubert varieties, flag manifolds. | Algebraic geometry – Projective and enumerative geometry – Configurations and arrangements of linear subspaces. | Nonassociative rings and algebras – Lie algebras and Lie superalgebras – Poisson algebras. | Several complex variables and analytic spaces – Singularities – Relations with arrangements of hyperplanes. | Algebraic topology – Homology and cohomology theories – Singular theory. | Algebraic topology – Fiber spaces and bundles – Discriminantal varieties, configuration spaces. | Algebraic topology – Applied homological algebra and category theory – Abstract complexes.

Classification: LCC QA564 .C663 2023 — DDC 516.3/5–dc23/eng20230722

LC record available at https://lccn.loc.gov/2023014589

---

10 9 8 7 6 5 4 3 2 1   23 22 21 20 19 18

Contents

Preface

Some of the most active and fruitful mathematical research occurs at the interface of algebraic geometry, representation theory, and topology. Noteworthy examples include the study of compactifications in three specific settings — algebraic group actions, configuration spaces, and hyperplane arrangements. These three types of compactifications enjoy common structural features, including relations to root systems, combinatorial descriptions of cohomology rings, the appearance of iterated blow-ups, the geometry of normal crossing divisors, and connections to mirror symmetry in physics. On the other hand, these compactifications are often studied independently of one another.

In the interest of deepening existing connections and forging new ones, we organized a conference called *Compactifications, configurations, and cohomology*. This fully in-person event was held on the campus of Northeastern University in Boston from October 22–24, 2021. It featured nine speakers at various career stages, who represented the full spectrum of themes from the previous paragraph. The following is a list of our speakers and their presentation titles; please note the slight discrepancy between our lists of speakers and authors.

- Ana Bălibanu (Harvard University): *Steinberg slices and group-valued moment maps*
- Patrick Brosnan (University of Maryland at College Park): *Fixed points in toroidal compactifications and essential dimension of covers*
- Graham Denham (Western University): *Bipermutohedral combinatorics*
- Nir Gadish (University of Michigan at Ann Arbor): *From compactified configurations on graphs to top weight cohomology of the moduli spaces* $M_{2,n}$
- Iva Halacheva (Northeastern University): *Cacti, De Concini–Procesi's wonderful compactification, and maximal commutative algebras*
- Ben Knudsen (Northeastern University): *Extremal stability for configuration spaces*
- Laurenţiu Maxim (University of Wisconsin at Madison): *On the topology of aspherical complex projective manifolds*
- Eric Ramos (Bowdoin College): *The categorical graph minor theorem and graph configuration spaces*
- Botong Wang (University of Wisconsin at Madison): *Singular Hodge theory of matroids*

This volume of Contemporary Mathematics is devoted to the research and exposition arising from our conference. It thereby focuses on new and existing connections between the aforementioned three types of compactifications, and in the process sets the stage for further research. It draws on the discipline-specific expertise of all authors, and at the same time give a unified, self-contained reference for compactifications and related constructions in different contexts.

We gratefully acknowledge our conference participants, as well as the authors of the articles appearing in this volume. Our conference was funded by the National Science Foundation, and hosted by the Department of Mathematics and College of Science at Northeastern University.

Peter Crooks
Alexandru I. Suciu

Contemporary Mathematics
Volume **790**, 2023
https://doi.org/10.1090/conm/790/15855

A quasi-Poisson structure on the multiplicative Grothendieck–Springer resolution

Ana Bălibanu

ABSTRACT. In this note we show that the multiplicative Grothendieck–Springer space has a natural quasi-Poisson structure. The associated group-valued moment map is the resolution morphism, and the quasi-Hamiltonian leaves are the connected components of the preimages of Steinberg fibers. This is a multiplicative analogue of the standard Poisson structure on the additive Grothendieck–Springer resolution, and an explicit illustration of a more general procedure of reduction along Dirac realizations which was developed in recent work of the author and Mayrand.

Introduction

The multiplicative Grothendieck–Springer resolution of a complex reductive group G is the incidence variety

$$\widetilde{G} = \{(g, B) \in G \times \mathcal{B} \mid g \in B\},$$

where $\mathcal{B}$ is the flag variety of G, viewed as the space of all Borel subgroups or, equivalently, of all Borel subalgebras. The first projection

$$\mu : \widetilde{G} \longrightarrow G$$

is a generically finite map which restricts to a resolution of singularities along each Steinberg fiber, and its restriction to the unipotent cone was originally constructed by Springer [**13**].

The group G is an example of a quasi-Poisson manifold. Such manifolds are generalizations of Poisson structures in which the Jacobi identity is twisted by a canonical trivector field, and they were introduced by Alekseev, Kosmann-Schwarzbach, and Meinrenken [**2**]. They are equipped with group-valued moment maps and are foliated by nondegenerate leaves, each of which carries a quasi-Hamiltonian 2-form. In particular, the quasi-Hamiltonian leaves of G are the conjugacy classes, which vary in dimension. Therefore, the quasi-Poisson structure on G is singular in the sense of foliation theory.

In this note we show that the Grothendieck–Springer resolution $\widetilde{G}$ also has a natural quasi-Poisson structure, whose group-valued moment map is μ. The leaves of this structure are the connected components of the preimages of Steinberg

---

2020 *Mathematics Subject Classification.* Primary 53D17; Secondary 20G05, 22E46.
This work was partially supported by a National Science Foundation MSPRF under award DMS–1902921.

fibers under μ, and they form a regular foliation of $\widetilde{G}$. The map μ is therefore a desingularization in two ways—it resolves each singular Steinberg fiber to a smooth variety, and it resolves the singular quasi-Hamiltonian foliation of G to a regular quasi-Hamiltonian foliation of $\widetilde{G}$.

The additive Grothendieck–Springer resolution. These observations are a multiplicative analogue of the Grothendieck–Springer resolution of the Lie algebra $\mathfrak{g}$ of G, which is the variety of pairs

$$\widetilde{\mathfrak{g}} = \{(x, \mathfrak{b}) \in \mathfrak{g} \times \mathcal{B} \mid x \in \mathfrak{b}\}$$

consisting of an element in the Lie algebra and a Borel subalgebra that contains it. We briefly recall its geometry, and we refer to [**10**] for more details. Fixing a Borel subgroup B of G and writing $\mathfrak{b}$ for its Lie algebra, there is a natural isomorphism

$$\begin{aligned} G \times_B \mathfrak{b} &\xrightarrow{\sim} \widetilde{\mathfrak{g}} \\ [g : x] &\longmapsto (\mathrm{Ad}_g(x), \mathrm{Ad}_g(\mathfrak{b})) \end{aligned}$$

which makes $\widetilde{\mathfrak{g}}$ into a vector bundle over G/B.

Let $\mathfrak{t}$ be a fixed maximal Cartan in $\mathfrak{b}$ with corresponding subgroup $T \subset B$, and let W be its Weyl group. Decompose $\mathfrak{b} = \mathfrak{t} \oplus \mathfrak{u}$, where $\mathfrak{u} = [\mathfrak{b}, \mathfrak{b}]$ is the nilpotent radical of $\mathfrak{b}$. The resolution $\widetilde{\mathfrak{g}}$ fits into the diagram

$$\begin{array}{ccccc} & & \widetilde{\mathfrak{g}} & & \\ & \swarrow^{\mu} & & \searrow^{\lambda} & \\ \mathfrak{g} & & & & \mathfrak{t} \\ & \searrow_{\kappa} & & \swarrow & \\ & & \mathfrak{t}/W, & & \end{array}$$

where μ is the first projection, λ is the projection of the bundle $G \times_B \mathfrak{b}$ onto the $\mathfrak{t}$-component of the fiber $\mathfrak{b}$, and κ is the composition of the adjoint quotient map with the Chevalley isomorphism

$$\mathfrak{g}/\!\!/G \cong \mathfrak{t}/W.$$

For any $s \in \mathfrak{t}$, write $\bar{s}$ for its image in $\mathfrak{t}/W$. When s is not regular, $\kappa^{-1}(\bar{s})$ is a singular algebraic variety and the map μ restricts to a resolution of singularities

$$G \times_B (s + \mathfrak{u}) \cong \lambda^{-1}(s) \longrightarrow \kappa^{-1}(\bar{s}).$$

In particular, in the special case when $s = 0$ this map is the Springer resolution

$$G \times_B \mathfrak{u} \longrightarrow \mathcal{N}$$

of the nilpotent cone $\mathcal{N}$ of $\mathfrak{g}$.

One can equip $\widetilde{\mathfrak{g}}$ with a Poisson structure through Hamiltonian reduction as follows. The action of G on itself by right multiplication induces Hamiltonian actions of G and B on the cotangent bundle T^*_G. The corresponding moment maps fit into the commutative diagram

$$\begin{array}{ccc} T^*_G \cong G \times \mathfrak{g} & & \\ & \overset{\nu}{\searrow} & \\ \Big\downarrow \overline{\nu} & & \mathfrak{g} \\ & \swarrow & \\ \mathfrak{g}/\mathfrak{u}. & & \end{array}$$

Here we use the left trivialization of the cotangent bundle and we identify $\mathfrak{g}^*$ with $\mathfrak{g}$ and $\mathfrak{b}^*$ with $\mathfrak{g}/\mathfrak{u}$ through the Killing form, so that the moment map ν is just the second projection. Since each point in the subset $\mathfrak{b}/\mathfrak{u} \subset \mathfrak{g}/\mathfrak{u}$ is fixed under the action of $\mathfrak{b}$ on $\mathfrak{g}/\mathfrak{u}$, the reduced space

$$\overline{\nu}^{-1}(\mathfrak{b}/\mathfrak{u})/B = \nu^{-1}(\mathfrak{b})/B = G \times_B \mathfrak{b}$$

inherits a natural Poisson structure from the canonical symplectic structure on T^*_G. Its symplectic leaves are the individual symplectic reductions

$$\nu^{-1}(s + \mathfrak{u})/B = G \times_B (s + \mathfrak{u}) \quad \text{for } s \in \mathfrak{t},$$

each of which is an affine bundle over G/B. Therefore the Grothendieck–Springer resolution $\widetilde{\mathfrak{g}}$ is a regular Poisson variety, and each resolution of singularities

$$G \times_B (s + \mathfrak{u}) \longrightarrow \kappa^{-1}(s)$$

is a symplectic resolution.

The group-valued analogue. In the quasi-Poisson setting, the role of the cotangent bundle is played by the internal fusion double $\mathbb{D}_G = G \times G$. A priori there is no analogue of Hamiltonian reduction with respect to the action of the Borel subgroup, because the G-valued moment map of $\mathbb{D}_G$ does not descend to a moment map with values in B—in other words, a quasi-Poisson G-space is not generally quasi-Poisson for the action of a subgroup of G.

However, by work of Bursztyn and Crainic [**8**, **9**], quasi-Poisson manifolds are an example of a more general class of structures known as Dirac manifolds. In [**5**], the author and Mayrand give a general procedure for Dirac reduction along a certain class of maps known as Dirac realizations. In this note, we show explicitly how this theory specializes to the Grothendieck–Springer space $\widetilde{G}$, using only the basics of Dirac structures. We prove the following result, as Theorems 2.3 and 2.4.

THEOREM. *There is a natural quasi-Poisson structure on the multiplicative Grothendieck–Springer resolution*

$$\widetilde{G} \cong G \times_B B,$$

which is inherited from the quasi-Hamiltonian structure on $\mathbb{D}_G$ and for which the map μ is a group-valued moment map. Its quasi-Hamiltonian leaves are the twisted unipotent bundles

$$G \times_B tU \quad \textit{for } t \in T.$$

The fiber bundle $G \times_B tU$ resolves the singularities of the Steinberg fiber F_t, which is a possibly singular quasi-Hamiltonian variety. This resolution is quasi-Hamiltonian, in the sense that the pullback of the corresponding 2-form along the desingularization map

$$G \times_B tU \longrightarrow F_t$$

agrees with the quasi-Hamiltonian structure on the leaf $G \times_B tU$. This quasi-Hamiltonian structure first appeared in work of Boalch [**6**, Section 4] in the setting of moduli spaces, and the quasi-Poisson structure on $\widetilde{G}$ which induces it can also be constructed using results of of Li-Bland and Ševera [**12**, Theorem 5].

We remark that the resolution $\widetilde{G}$ also carries a Poisson structure, which was introduced by Evens and Lu [**11**] and which is obtained from a Poisson structure on the double $\mathbb{D}_G$ through coisotropic reduction. In this work, the authors equip G with a compatible Poisson structure that can be viewed as the semi-classical limit of the quantum group $U_q(\mathfrak{g})$, and with respect to which the resolution map μ is Poisson. The T-orbits of the symplectic leaves of G are intersections of conjugacy classes and Bruhat cells, and the singularities of their closures are resolved by the T-orbits of symplectic leaves in $\widetilde{G}$, which are regular Poisson manifolds.

1. Dirac manifolds and quasi-Poisson structures

1.1. Dirac structures. We begin by giving some background on Dirac manifolds, and we refer the reader to [**7**] for more details. Let M be a (real or complex) manifold and let $\eta \in \Omega^3(M)$ be a closed 3-form. A *η-twisted Dirac structure* on M is a subbundle $L \subset T_M \oplus T_M^*$ such that

- L is Lagrangian with respect to the nondegenerate symmetric pairing
$$\langle (X, \alpha), (Y, \beta) \rangle = \alpha(Y) + \beta(X), \text{ and}$$
- the space of sections $\Gamma(L)$ is closed under the η-twisted Dorfman bracket
$$[\![(X, \alpha), (Y, \beta)]\!] = ([X, Y], \mathcal{L}_X\beta + \iota_Y\alpha + \iota_{X\wedge Y}\eta).$$

While the Dorfman bracket is not a Lie bracket on $T_M \oplus T_M^*$, it restricts to a Lie bracket on the sections of the subbundle L. In this way L becomes a Lie algebroid over M, with anchor map the first projection

$$\begin{aligned} p_T : \quad L \quad &\longrightarrow T_M \\ (X, \alpha) &\longmapsto X. \end{aligned}$$

EXAMPLE 1.1. Let $\omega \in \Omega^2(M)$ be any 2-form, and let $\eta = -d\omega$. Then the graph

$$L_\omega = \{(X, \omega^\flat(X)) \mid X \in T_M\}$$

is a η-twisted Dirac structure on M. Such Dirac structures are called *nondegenerate* or *presymplectic*, and are characterized by the property

$$L_\omega \cap T_M^* = 0.$$

Note that here and throughout we view T_M and T_M^* as subbundles of $T_M \oplus T_M^*$ via the coordinate embeddings. In particular, a Dirac structure L on M is induced by a symplectic form if and only if it is non-twisted and

$$L \cap T_M^* = L \cap T_M = 0.$$

More generally, if (M, L) is a η-twisted Dirac manifold, the image of the anchor map p_T is an involutive generalized distribution which integrates to a foliation of M by *nondegenerate* or *presymplectic leaves.* Each leaf S carries a natural presymplectic form ω_S, which has the property that

$$d\omega_S = -\eta_{|S}.$$

EXAMPLE 1.2. Let $\pi \in \mathcal{X}^2(M)$ be a bivector on M and suppose that there is a 3-form $\eta \in \Omega^3(M)$ such that

$$[\pi, \pi] = 2\pi^{\#}(\eta),$$

where the left-hand side denotes the Schouten–Nijenhuis bracket. Then π is a *twisted Poisson structure*, and its graph

$$L_\pi = \{(\pi^{\#}(\alpha), \alpha) \mid \alpha \in T_M^*\}$$

is a η-twisted Dirac structure on M. Conversely, a Dirac structure L on M is induced by a Poisson bivector if and only it is non-twisted and

$$L \cap T_M = 0.$$

In this case, the foliation of M by nondegenerate leaves is exactly the symplectic foliation.

Given a map $f : M \longrightarrow N$ and a η_N-twisted Dirac structure L_N on N, the *pullback* of L_N under f is the generalized distribution

$$f^*L_N = \{(X, f^*\alpha) \in T_M \oplus T_M^* \mid (f_*X, \alpha) \in L_N\}.$$

If f^*L_N is a smooth bundle, it defines a $f^*\eta_N$-twisted Dirac structure on M [**7**, Proposition 1.10]. When M is equipped with a twisted Dirac structure L_M, the map f is called *backward-Dirac* (or *b-Dirac*) if

$$L_M = f^*L_N.$$

Such maps generalize the pullbacks of differential forms. In particular, when (M, ω_M) and (N, ω_N) are nondegenerate Dirac manifolds, f is b-Dirac if and only if

$$f^*\omega_N = \omega_M.$$

Similarly, if L_M is a η_M-twisted Dirac structure on M, the *pushforward* of L_M under f is the generalized distribution

$$f_*L_M = \{(f_*X, \alpha) \in T_N \oplus T_N^* \mid (X, f^*\alpha) \in L_M\},$$

as long as it is well-defined. When N is equipped with a twisted Dirac structure L_N, the map f is called *forward-Dirac* (or *f-Dirac*) if at every point it satisfies

$$L_N = f_*L_M.$$

Such maps generalize pushforwards of vector fields. In particular, when (M, π_M) and (N, π_N) are Poisson manifolds, f is f-Dirac if and only if

$$f_*\pi_M = \pi_N.$$

When $f : M \longrightarrow N$ is a diffeomorphism, it is f-Dirac if and only if it is b-Dirac, and in this case it is called a *Dirac diffeomorphism*. Suppose now that a group G acts on M by Dirac diffeomorphisms and that M/G has the structure of a manifold such that the quotient map

$$q : M \longrightarrow M/G$$

is a smooth submersion. Write $\mathfrak{g}$ for the Lie algebra of G and

$$\begin{aligned} \rho_M : \mathfrak{g} &\longrightarrow \mathcal{X}(M) \\ \xi &\longmapsto \xi_M \end{aligned}$$

for the infinitesimal action map. In this case the pushforward $f_* L_M$ is a Dirac structure on M/G if the action of G on M is *regular*—that is, if the generalized distribution

$$\rho_M(\mathfrak{g}) \cap L_M \subset T_M$$

has constant dimension [**7**, Proposition 1.13].

1.2. Quasi-Poisson manifolds. Let G be a (real or complex) Lie group whose Lie algebra $\mathfrak{g}$ carries an invariant, nondegenerate, symmetric bilinear form $(\cdot, \cdot)$. The *Cartan 3-form* of G is the invariant 3-form induced by the element $\eta \in \wedge^3 \mathfrak{g}^*$ defined by

$$\eta(x, y, z) = \frac{1}{12}(x, [y, z]) \quad \text{for all } x, y, z \in \mathfrak{g},$$

and we denote by $\chi \in \wedge^3 \mathfrak{g}$ the 3-tensor which corresponds to it under the induced identification $\mathfrak{g} \cong \mathfrak{g}^*$.

Through the infinitesimal action map, χ generates a canonical bi-invariant trivector field

$$\chi_M \in \mathcal{X}^3(M).$$

A *quasi-Poisson* structure on M is a bivector field $\pi \in \mathcal{X}^2(M)$ whose Schouten bracket satisfies

$$[\pi, \pi] = \chi_M.$$

This notion was first introduced in a series of papers by Alekseev, Malkin, and Meinrenken [**3**], by Alekseev and Kosmann-Schwarzbach [**1**], and by Alekseev, Kosmann-Schwarzbach, and Meinrenken [**2**]. Viewed as a skew-symmetric bracket on the space of functions on M, a quasi-Poisson structure is a biderivation which satisfies a χ_M-twisted version of the Jacobi identity. In particular, when the action of G on M is trivial, the Cartan trivector field χ_M vanishes and we recover the usual definition of a Poisson manifold.

Consider the map

$$\begin{aligned} \sigma : \mathfrak{g} &\longrightarrow T_G^* \\ \xi &\longmapsto \frac{1}{2}(\xi^R + \xi^L)^\vee, \end{aligned}$$

where ξ^R and ξ^L are the right- and left-invariant vector fields induced by the Lie algebra element ξ, and $v^\vee \in T_G^*$ is the 1-form corresponding to the vector field $v \in T_G$ under the isomorphism given by left-invariant bilinear form induced by $(\cdot, \cdot)$ on T_G. We let

$$\sigma^\vee : T_G^* \longrightarrow \mathfrak{g}$$

be its adjoint. The quasi-Poisson manifold (M, π) is *Hamiltonian* if it is equipped with a G-equivariant map

$$\Phi : M \longrightarrow G,$$

called a *group-valued moment map*, which satisfies the condition

$$\pi^\# \circ \Phi^* = \rho_M \circ \sigma^\vee. \tag{1.1}$$

The Hamiltonian quasi-Poisson manifold (M, π, Φ) is *nondegenerate* if the bundle map

$$\begin{aligned} T_M^* \oplus \mathfrak{g} &\longrightarrow T_M \\ (\alpha, \xi) &\longmapsto \pi^\#(\alpha) + \xi_M \end{aligned} \tag{1.2}$$

is surjective. In this case, M carries a *quasi-Hamiltonian* 2-form $\omega \in \Omega^2(M)$ which satisfies the compatibility condition

$$\pi^{\#} \circ \omega^{\flat} = C, \tag{1.3}$$

where

$$C := 1 - \frac{1}{4}\rho_M \circ \rho^{\vee} \circ \Phi_*$$

and

$$\rho^{\vee} : T_G \longrightarrow \mathfrak{g}$$

is the adjoint of the infinitesimal action of G on itself by conjugation [**2**, Lemma 10.2]. This 2-form has the property that

$$d\omega = -\Phi^*\eta,$$

and the moment map condition (1.1) can be rewritten as

$$\omega^{\flat} \circ \rho_M = \Phi^* \circ \sigma. \tag{1.4}$$

Moreover, if (1.2) fails to be surjective, its image is an integrable generalized distribution and the quasi-Poisson manifold M is foliated by quasi-Hamiltonian leaves.

1.3. Quasi-Poisson bivectors as Dirac structures. The quasi-Poisson structure on the G-manifold (M, π, Φ) is encoded [**8**, Theorem 3.16] by the $\Phi^*\eta$-twisted Dirac bundle

$$L = \{(\pi^{\#}(\alpha) + \xi_M, C^*(\alpha) + \Phi^*\sigma(\xi)) \mid \alpha \in T^*_M,\, \xi \in \mathfrak{g}\}.$$

The image of its projection onto the tangent component is the generalized distribution given by the map (1.2), and the nondegenerate leaves associated to this Dirac structure are precisely the quasi-Hamiltonian leaves. In particular, if M is quasi-Hamiltonian with 2-form $\omega \in \Omega^2(M)$, by (1.3) and (1.4) this bundle can be writen

$$L = \{(X, \omega^{\flat}(X)) \mid X \in T_M\}.$$

EXAMPLE 1.3 ([**2**, Proposition 3.1]). The group G has a natural quasi-Poisson structure relative to the conjugation action, called the *Cartan–Dirac structure*, whose moment map is the identity. The associated Dirac bundle [**8**, Example 3.4] is

$$L_G = \{(\rho(\xi), \sigma(\xi)) \mid \xi \in \mathfrak{g}\} = \left\{(\xi^L - \xi^R, \sigma(\xi)) \mid \xi \in \mathfrak{g}\right\}. \tag{1.5}$$

Projecting onto the tangent component, we see that the nondegenerate leaves of this structure are the conjugacy classes. Therefore the Cartan–Dirac structure on G can be seen as a multiplicative analogue of the classical Kirillov–Kostant–Souriau Poisson structure on $\mathfrak{g}^* \cong \mathfrak{g}$.

EXAMPLE 1.4 ([**2**, Example 5.3]). The *internal fusion double* $\mathbb{D}_G := G \times G$ has a natural nondegenerate quasi-Poisson structure relative to the $G \times G$-action

$$(g_1, g_2)(a, b) = (g_1 a g_2^{-1}, g_2 b g_2^{-1}).$$

This structure is a multiplicative counterpart of the canonical symplectic structure on the cotangent bundle of G, viewed under the identification

$$T^*_G \cong G \times \mathfrak{g}$$

induced by left-trivialization and by the invariant inner product. Its moment map is given by

$$\Phi: \mathbb{D}_G \longmapsto G \times G \tag{1.6}$$
$$(a,b) \longmapsto (aba^{-1}, b^{-1}).$$

Viewing the quasi-Poisson G-manifold (M, π, Φ) as a Dirac manifold with $\Phi^*\eta$-twisted Dirac structure L, the moment map Φ is a f-Dirac map which satisfies the additional nondegeneracy condition

$$\ker \Phi_* \cap L = 0. \tag{1.7}$$

In fact, this property completely characterizes Hamiltonian quasi-Poisson manifolds [**8**, Proposition 3.20]—suppose that (M, L) is a Dirac manifold and that

$$\Phi : (M, L) \longrightarrow (G, L_G)$$

is a f-Dirac map which satisfies (1.7). Then, for any $(\rho(\xi), \sigma(\xi)) \in L_G$, there is a unique pair $(X_\xi, \alpha_\xi) \in L$ such that

$$\Phi_* X_\xi = \rho(\xi) \qquad \text{and} \qquad \Phi^*\sigma(\xi) = \alpha_\xi.$$

This gives an infinitesimal action of the Lie algebra $\mathfrak{g}$ on M via

$$\rho_M : \mathfrak{g} \longrightarrow \mathcal{X}(M) \tag{1.8}$$
$$\xi \longmapsto X_\xi.$$

Moreover, for any $\alpha \in T^*M$, there is a unique $X_\alpha \in T_M$ such that

$$\Phi_* X_\alpha = \sigma^{\vee *} \rho_M^*(\alpha) \qquad \text{and} \qquad (X_\alpha, C^*(\alpha)) \in L.$$

The map

$$\pi^\# : T_M^* \longrightarrow T_M$$
$$\alpha \longmapsto X_\alpha$$

defines a quasi-Poisson bivector $\pi \in \mathcal{X}^2(M)$ relative to the action (1.8), whose associated moment map is Φ.

2. The multiplicative Grothendieck–Springer resolution

2.1. The Grothendieck–Springer space. From now on let G be a reductive complex group, so that the nondegenerate bilinear form on every simple factor of $\mathfrak{g}$ is given by a scalar multiple of the Killing form, and once again write $\mathcal{B}$ for the flag variety of all Borel subgroups. The *Grothendieck–Springer simultaneous resolution* of G is the variety of pairs

$$\widetilde{G} := \{(g', B') \in G \times \mathcal{B} \mid g' \in B'\}.$$

There is a natural map

$$\mu : \widetilde{G} \longrightarrow G \tag{2.1}$$

given by the first projection, whose general fiber is finite—when $g' \in G$ is regular and semisimple, the points of $\mu^{-1}(g')$ are permuted freely and transitively by W, the Weyl group of G.

Fixing a Borel subgroup B of G, the isomorphism

$$\begin{aligned} G \times_B B &\xrightarrow{\sim} \widetilde{G} \\ [g : b] &\longmapsto (gbg^{-1}, gBg^{-1}) \end{aligned}$$

realizes $\widetilde{G}$ as a fiber bundle over $G/B \cong \mathcal{B}$. Let T be a maximal torus of B and let $U = [B, B]$ be its unipotent radical, so that we have a splitting $B = TU$. Since B stabilizes each coset tU, there is a well-defined map

$$\begin{aligned} \lambda : G \times_B B &\longrightarrow T \\ [g : tu] &\longmapsto t \end{aligned}$$

whose fibers

$$\lambda^{-1}(t) = G \times_B tU$$

are multiplicative analogues of twisted cotangent bundles over G/B.

Let $G/\!\!/G = \operatorname{Spec} \mathbb{C}[G]^G$ be the adjoint quotient of G. By the Chevalley theorem, the restriction map gives an isomorphism of algebras

$$\mathbb{C}[G]^G \cong \mathbb{C}[T]^W$$

and therefore an identification of varieties

$$G/\!\!/G \cong T/W.$$

The two quotient maps

$$G \longrightarrow G/\!\!/G \qquad \text{and} \qquad T \longrightarrow T/W$$

fit into a commutative diagram

(2.2)
$$\begin{array}{ccccc} & & \widetilde{G} & & \\ & \swarrow^{\mu} & & \searrow^{\lambda} & \\ G & & & & T \\ & \searrow_{\kappa} & & \swarrow & \\ & & T/W & & \end{array}$$

whose restriction to the regular locus of G is Cartesian.

For any $t \in T$, let $\bar{t}$ be its image in T/W. The fibers of κ, which are irreducible subvarieties of G of codimension equal to the rank, are called *Steinberg fibers*, and each is a union of finitely many conjugacy classes. In particular, the Steinberg fiber

$$F_t := \kappa^{-1}(\bar{t})$$

contains a unique regular conjugacy class, which is open and dense, and the unique semisimple conjugacy class $G \cdot t$, which is closed and of minimal dimension [**14**, Theorem 6.11 and Remark 6.15]. In particular, if $t \in T$ is regular, then the fiber F_t is simply the orbit $G \cdot t$ and therefore it is a smooth variety isomorphic to the quotient G/T.

When $t \in T$ is not regular, the Steinberg fiber F_t is generally singular. Diagram (2.2) gives a natural surjection $\lambda^{-1}(t) \longrightarrow \kappa^{-1}(\bar{t})$, and we obtain a proper birational map

$$G \times_B tU \longrightarrow F_t$$

which is a resolution of singularities. In the special case of the identity element $1 \in T$, the Steinberg fiber $\mathcal{U} := F_1$ is the variety of all unipotent elements of G, and this birational map is precisely the Springer resolution

$$G \times_B U \longrightarrow \mathcal{U}.$$

2.2. A quasi-Poisson structure on $\widetilde{G}$. In this section we will show that $G \times_B B$ carries a natural quasi-Poisson structure for the left action of G given by

$$h \cdot [g : b] = [hg : b]. \tag{2.3}$$

Consider the diagram

$$\begin{array}{ccccccc} & & G \times B & & \overset{\jmath}{\hookrightarrow} & & \mathbb{D}_G \\ & \swarrow^{q} & & \searrow^{\Phi} & & & \searrow^{\Phi} \\ G \times_B B & & & & G \times B & \overset{\imath}{\hookrightarrow} & G \times G \\ & \searrow_{\mu} & & \swarrow_{p} & & & \\ & & G, & & & & \end{array} \tag{2.4}$$

where p and q are the natural quotient maps, μ is the map defined in (2.1), and Φ is the quasi-Poisson moment map of the internal fusion double given by (1.6).

The bundle $\jmath^* L_{\mathbb{D}_G}$ is the graph of the two-form $\jmath^*\omega$, and is therefore a Dirac structure on $G \times B$. To check that it descends to the desired Dirac structure on the Grothendieck–Springer resolution, we will begin with the following simple lemma.

LEMMA 2.1. *Let $\xi \in \mathfrak{b}$. Then $T_B \subset \ker \sigma(\xi)$ if and only if $\xi \in \mathfrak{u}$.*

PROOF. Fix a point $tu \in B$, with semisimple part $t \in T$ and unipotent part $u \in U$, and write $\xi = s + n \in \mathfrak{t} \oplus \mathfrak{u}$. Any vector in $T_{tu}B$ is of the form x^R for some $x \in \mathfrak{b}$, and we have

$$\begin{aligned} x^R \in \ker(\xi^R + \xi^L)^\vee \quad \text{for all } x \in \mathfrak{b} \quad &\Longleftrightarrow \quad (\xi + \mathrm{Ad}_{tu}\,\xi, x) = 0 \quad \text{for all } x \in \mathfrak{b} \\ &\Longleftrightarrow \quad \xi + \mathrm{Ad}_{tu}\,\xi \in \mathfrak{u} \\ &\Longleftrightarrow \quad s = 0, \end{aligned}$$

where the second equivalence follows from the fact that $\mathfrak{b}^\perp = \mathfrak{u}$ under the Killing form. □

PROPOSITION 2.2. *The action of B on $G \times B$ defined by*

$$h \cdot (g, b) = (gh^{-1}, hbh^{-1})$$

for $h \in B$ and $(g, b) \in G \times B$ is a regular Dirac action with respect to the Dirac structure $\jmath^* L_{\mathbb{D}_G}$.

PROOF. Since the second copy of G acts on $\mathbb{D}_G$ by Dirac automorphisms, the same is true for the action of B on the submanifold $G \times B$. It remains only to show that this action is regular.

We have

$$\begin{aligned} \rho_{\mathbb{D}_G}(0 \oplus \mathfrak{b}) \cap \jmath^* L_{\mathbb{D}_G} &= \{\rho_{\mathbb{D}_G}(0, \xi) \mid \xi \in \mathfrak{b} \text{ and } T_{G \times B} \subset \ker \Phi^* \sigma(0, \xi)\} \\ &= \{\rho_{\mathbb{D}_G}(0, \xi) \mid \xi \in \mathfrak{b} \text{ and } T_{G \times B} \subset \ker \sigma(0, \xi)\} \\ &= \{\rho_{\mathbb{D}_G}(0, \xi) \mid \xi \in \mathfrak{u}\}, \end{aligned}$$

where the last equality follows from Lemma 2.1. Since the action of B on $G\times B$ has trivial stabilizers, this implies that the intersection $\rho_{\mathbb{D}_G}(0\oplus\mathfrak{b})\cap\jmath^*L_{\mathbb{D}_G}$ has constant rank and therefore that this action is regular. □

Because the Borel subalgebra satisfies

$$\mathfrak{b}^\perp = [\mathfrak{b},\mathfrak{b}],$$

the restriction of the Cartan 3-form η to the subgroup B vanishes. Chasing diagram (2.4), we see that

$$\begin{aligned} q^*\mu^*\eta &= \Phi^*p^*\eta \\ &= \Phi^*\imath^*(\eta_1,\eta_2) \\ &= \jmath^*\Phi^*(\eta_1,\eta_2), \end{aligned}$$

so the twist of the Dirac structure $\jmath^*L_{\mathbb{D}_G}$ is in the image of q^*. By [**7**, Proposition 1.13], Proposition 2.2 then implies that the pushforward

$$L_{\widetilde{G}} := q_*\jmath^*L_{\mathbb{D}_G}$$

is a smooth vector bundle and therefore a $\mu^*\eta$-twisted Dirac structure on $\widetilde{G}$. We now show that it corresponds to a quasi-Poisson bivector.

THEOREM 2.3. *The Dirac structure $L_{\widetilde{G}}$ is induced by a quasi-Poisson structure on $G\times_B B$ whose moment map is μ.*

PROOF. To show that $L_{\widetilde{G}}$ defines a quasi-Poisson structure with moment map μ, we must show (i) that μ is a f-Dirac map which satisfies the nondegeneracy condition (1.7), and (ii) that the induced G-action on $G\times_B B$ coincides with the action given in (2.3).

(i) To see that μ is f-Dirac, we use diagram (2.4) to compute

$$\begin{aligned} L_G &= p_*\imath^*\Phi_*L_{\mathbb{D}_G} \\ &= p_*\Phi_*\jmath^*L_{\mathbb{D}_G} \\ &= \mu_*q_*\jmath^*L_{\mathbb{D}_G} = \mu_*L_{\widetilde{G}}, \end{aligned}$$

where the second equality follows from [**4**, Lemma 1.19]. Checking the nondegeneracy condition

$$\ker\mu_*\cap L_{\widetilde{G}} = 0,$$

is equivalent to showing that

$$\ker(p_*\Phi_*)\cap\jmath^*L_{\mathbb{D}_G}\subset\rho_{\mathbb{D}_G}(0\oplus\mathfrak{b}). \tag{2.5}$$

Suppose that $(X,0)\in\jmath^*L_{\mathbb{D}_G}$ satisfies $p_*\Phi_*(X)=0$. Then there is a 1-form $\alpha\in T^\circ_{G\times B}$ such that

$$(X,\alpha)\in L_{\mathbb{D}_G}.$$

Moreover, since $T^\circ_{G\times B}=\Phi^*T^\circ_{G\times B}$ and since Φ is f-Dirac, there is a further $\beta\in T^\circ_{G\times B}$ such that (Φ_*X,β) is an element of the Cartan–Dirac structure $L_{G\times G}$. Since $p_*\Phi_*X=0$ and since X is tangent to $G\times B$, we obtain

$$\Phi_*X=\rho(0,\xi)$$

for some $\xi\in\mathfrak{b}$. By (1.5), this means that $\beta=\sigma(0,\xi)$ and we get

$$(X,\Phi^*\beta)\in L_{\mathbb{D}_G}\qquad\text{and}\qquad(\rho_{\mathbb{D}_G}(0,\xi),\Phi^*\beta)\in L_{\mathbb{D}_G}.$$

Since $\Phi_* X = \Phi_* \rho_{\mathbb{D}_G}(0,\xi)$, the nondegeneracy condition (1.2) applied to the moment map Φ implies that $X = \rho_{\mathbb{D}_G}(0,\xi)$, proving (2.5).

It remains to show that the infinitesimal action of $\mathfrak{g}$ on $G \times_B B$ induced by the map μ agrees with the action defined in (2.3). For this, let $\xi \in \mathfrak{g}$ and notice that, since $L_{\mathbb{D}_G}$ is quasi-Poisson for the action of $G \times G$ on $\mathbb{D}_G$, we have

$$(\rho_{\mathbb{D}_G}(\xi,0), \Phi^*\sigma(\xi,0)) \in L_{\mathbb{D}_G}.$$

Since $\rho_{\mathbb{D}_G}(\xi,0)$ is tangent to $G \times B$, it follows that

$$(\rho_{\mathbb{D}_G}(\xi,0), \jmath^*\Phi^*\sigma(\xi,0)) \in L_{G\times B}.$$

Moreover, by chasing diagram (2.4) we see that

$$\jmath^*\Phi^*\sigma(\xi,0) = \Phi^*\imath^*\sigma(\xi,0) = \Phi^*p^*\sigma(\xi) = q^*\mu^*\sigma(\xi),$$

and we get

$$(q_*\rho_{\mathbb{D}_G}(\xi,0), \mu^*\sigma(\xi)) \in L_{\widetilde{G}}.$$

Since $\mu_* q_* \rho_{\mathbb{D}_G}(0,\xi) = \rho(0,\xi)$, by (1.8) this implies that ξ acts on $G \times_B B$ by the vector field $q_*\rho_{\mathbb{D}_G}(\xi,0)$, which is precisely the vector field corresponding to ξ under the action (2.3). □

THEOREM 2.4. *The quasi-Hamiltonian leaves of* $L_{\widetilde{G}}$ *are the twisted unipotent bundles*

$$G \times_B tU \quad \text{for } t \in T.$$

PROOF. Let $X \in T_{G\times B}$. The pushforward q_*X is contained in $p_T(L_{\widetilde{G}})$ if and only if

$$\begin{aligned} \jmath^*\omega^\flat(X) \in \operatorname{im} q^* \quad &\Longleftrightarrow \quad \omega(X, \rho_{\mathbb{D}_G}(0,\xi)) = 0 \quad \text{for all } \xi \in \mathfrak{b} \\ &\Longleftrightarrow \quad X \in \ker \Phi^*\sigma(0,\xi) \quad \text{for all } \xi \in \mathfrak{b} \\ &\Longleftrightarrow \quad \Phi_*X \in \ker \sigma(0,\xi) \quad \text{for all } \xi \in \mathfrak{b}, \end{aligned} \tag{2.6}$$

where the second equivalence follows from the moment map condition (1.4).

Keeping the notation of Lemma 2.1 and letting $x \in \mathfrak{b}$, we have

$$\begin{aligned} x^R \in \ker(\xi^R + \xi^L)^\vee \text{ for all } \xi \in \mathfrak{b} \quad &\Longleftrightarrow \quad (\xi + \operatorname{Ad}_{tu}\xi, x) = 0 \text{ for all } \xi \in \mathfrak{b} \\ &\Longleftrightarrow \quad (\xi, x + \operatorname{Ad}_{tu}^{-1} x) = 0 \text{ for all } \xi \in \mathfrak{b} \\ &\Longleftrightarrow \quad x \in \mathfrak{u}, \end{aligned} \tag{2.7}$$

where the last equivalence follows from Lemma 2.1.

Using (2.6) and (2.7),

$$p_T(L_{\widetilde{G}}) = \{q_*X \mid \jmath^*\omega^\flat(X) \in \operatorname{im} q^*\} = \{q_*X \mid X \in T_{G\times tU}\},$$

and so the leaves integrating the generalized distribution $p_T(L_{\widetilde{G}})$ are precisely the submanifolds of $G \times_B B$ of the form

$$G \times_B tU \quad \text{for } t \in T. \qquad \square$$

References

[1] A. Alekseev and Y. Kosmann-Schwarzbach, *Manin pairs and moment maps*, J. Differential Geom. **56** (2000), no. 1, 133–165. MR1863024

[2] A. Alekseev, Y. Kosmann-Schwarzbach, and E. Meinrenken, *Quasi-Poisson manifolds*, Canad. J. Math. **54** (2002), no. 1, 3–29, DOI 10.4153/CJM-2002-001-5. MR1880957

[3] A. Alekseev, A. Malkin, and E. Meinrenken, *Lie group valued moment maps*, J. Differential Geom. **48** (1998), no. 3, 445–495. MR1638045

[4] A. Bălibanu, *Steinberg slices and group-valued moment maps*, Adv. Math. **402** (2022), Paper No. 108344, 46, DOI 10.1016/j.aim.2022.108344. MR4396508

[5] A. Balibanu and M. Mayrand, *Reduction of Dirac realizations*, `arXiv:2210.07200`.

[6] P. P. Boalch, *Riemann-Hilbert for tame complex parahoric connections*, Transform. Groups **16** (2011), no. 1, 27–50, DOI 10.1007/s00031-011-9121-1. MR2785493

[7] H. Bursztyn, *A brief introduction to Dirac manifolds*, Geometric and topological methods for quantum field theory, Cambridge Univ. Press, Cambridge, 2013, pp. 4–38. MR3098084

[8] H. Bursztyn and M. Crainic, *Dirac structures, momentum maps, and quasi-Poisson manifolds*, The breadth of symplectic and Poisson geometry, Progr. Math., vol. 232, Birkhäuser Boston, Boston, MA, 2005, pp. 1–40, DOI 10.1007/0-8176-4419-9_1. MR2103001

[9] H. Bursztyn and M. Crainic, *Dirac geometry, quasi-Poisson actions and D/G-valued moment maps*, J. Differential Geom. **82** (2009), no. 3, 501–566. MR2534987

[10] N. Chriss and V. Ginzburg, *Representation theory and complex geometry*, Modern Birkhäuser Classics, Birkhäuser Boston, Ltd., Boston, MA, 2010. Reprint of the 1997 edition, DOI 10.1007/978-0-8176-4938-8. MR2838836

[11] S. Evens and J.-H. Lu, *Poisson geometry of the Grothendieck resolution of a complex semisimple group* (English, with English and Russian summaries), Mosc. Math. J. **7** (2007), no. 4, 613–642, 766, DOI 10.17323/1609-4514-2007-7-4-613-642. MR2372206

[12] D. Li-Bland and P. Ševera, *Symplectic and Poisson geometry of the moduli spaces of flat connections over quilted surfaces*, Mathematical aspects of quantum field theories, Math. Phys. Stud., Springer, Cham, 2015, pp. 343–411. MR3330247

[13] T. A. Springer, *The unipotent variety of a semi-simple group*, Algebraic Geometry (Internat. Colloq., Tata Inst. Fund. Res., Bombay, 1968), Oxford Univ. Press, London, 1969, pp. 373–391. MR0263830

[14] R. Steinberg, *Regular elements of semisimple algebraic groups*, Inst. Hautes Études Sci. Publ. Math. **25** (1965), 49–80. MR180554

Department of Mathematics, Harvard University, 1 Oxford Street, Cambridge, Massachusetts 02138

Email address: `ana@math.harvard.edu`

Contemporary Mathematics
Volume **790**, 2023
https://doi.org/10.1090/conm/790/15856

Volumes of definable sets in o-minimal expansions and affine GAGA theorems

Patrick Brosnan

ABSTRACT. In this mostly expository note, I give a quick proof of the definable Chow theorem of Peterzil and Starchenko using the Bishop-Stoll theorem and a volume estimate for definable sets due to Nguyen and Valette. The volume estimate says that any d-dimensional definable subset of $S \subseteq \mathbb{R}^n$ in an o-minimal expansion of the ordered field of real numbers satisfies the inequality $\mathcal{H}^d(\{x \in S : \|x\| < r\}) \leq Cr^d$, where $\mathcal{H}^d$ denotes the d-dimensional Hausdorff measure on $\mathbb{R}^n$ and C is a constant depending on S. Since this note is intended to be helpful to algebraic geometers not versed in o-minimal structures and definable sets, I review these notions and also prove the main volume estimate from scratch.

1. Introduction

The GAGA theorem of Peterzil–Starchenko [**14**] says that a closed analytic subset of $\mathbb{C}^n$, which is definable in an o-minimal expansion of the ordered field $\mathbb{R}$, is algebraic. This theorem, which I restate as Theorem 5.3 below, is a crucial ingredient in (at least) two, closely related, recent advances in Hodge theory: the paper by Bakker, Klingler and Tsimerman [**1**], which gives a new proof of the theorem of Cattani, Deligne and Kaplan [**4**] on the algebraicity of the Hodge locus, and the paper by Bakker, Brunebarbe and Tsimerman proving a conjecture of Griffiths on the algebraicity of the image of the period map [**2**]. Given these important results, it seems desirable to have an understanding of the Peterzil–Starchenko theorem from several points of view.

The point of this (mainly expository) note is to show that the theorem of Peterzil–Starchenko follows from a Chow theorem originally due to Stoll [**17**] and a volume estimate for definable sets. As I will explain in §1.1, this volume estimate was known in various forms to experts for some time. I will state it precisely in Theorem 4.1 below and prove it from scratch in §6, but essentially it says the following: Suppose S is a d-dimensional subset of $\mathbb{R}^n$, which is definable with respect to an o-minimal expansion of $\mathbb{R}_{\mathrm{alg}}$ (for example, $\mathbb{R}_{\mathrm{an,exp}}$). Then the Hausdorff measure of the set $S(r) := \{x \in S : \|x\| < r\}$ viewed as a function of r is in $O(r^d)$. In other words, the volume of the intersection of S with a ball of radius r grows at most as fast as a constant multiple of r^d. In §5, I use the volume estimate to give a quick proof of the Peterzil-Starchenko theorem.

The author gratefully acknowledges support from a Simons Foundation Grant.

1.1. History. Peterzil and Starchenko gave two proofs of their theorem. The first, published in 2008 in [**14**], works for o-minimal expansions of arbitrary real closed fields and is based on results from model theory. The second proof, published in 2009 in [**15**], like the proof presented in this note, relies on results from complex analysis. More precisely, it relies on a paper of Shiffman that is related to Stoll's theorem [**16**] in that Shiffman's results are ultimately about bounds on volumes of complex analytic sets. It should be noted that the result in [**14**] is strictly stronger than the result proved in this paper (in Theorem 5.3 below) and the result in [**15**] because the hypotheses are much more general.

After the first version of this paper appeared on the ArXiv, I learned of several other proofs of my main volume estimate Theorem 4.1 as well as another short way of proving Theorem 5.3.

Firstly, as Xuan Viet Nhan Nguyen pointed out, Theorem 4.1 is a special case of Proposition 3.1 of a 2018 paper by Nguyen and Valette [**13**]. Moreover my proof of Theorem 4.1, which relies on the fact that definable sets in o-minimal structures have cell decompositions, is similar to the proof of [**13**, Proposition 3.1].

Secondly, as Tobias Kaiser pointed out, in the context of subanalytic sets, a result closely related to Theorem 4.1 follows easily from a 1989 paper by Kurdyka and Raby [**9**].

Finally, after I posted a version of this paper revised to include the references to the above works of, Georges Compte emailed to tell me that Theorem 4.1 and, in fact, even Theorem 5.3 can both be deduced from results in E. M. Chirka's book *Complex analytic sets* [**5**], which was published in 1985 in Russian and in 1989 in English. More specifically, it is possible to give a short proof of Theorem 5.3 using [**5**, Theorem 1, p. 243].

1.2. Outline. In spite of the above history, I feel that the viewpoint in this paper has value coming from its brevity and the directness of the connection between Stoll's theorem and Peterzil—Starchenko. I also hope that the paper will be readable by an audience not familiar with o-minimal models or with Stoll's theorem.

To help make this paper approachable for algebraic geometers, I review the theory of o-minimal structures in section §2. In §3, I review the notions of Hausdorff dimension and Stoll's theorem, which is also called the Bishop–Stoll theorem. (Bishop [**3**] generalized and extended the result of Stoll used in this paper.) In §4, I state the main volume estimate, Theorem 4.1, after a discussion of cells in §4.1. I think of cellular decompositions as being the superpower of definable sets in o-minimal structures, and, while I do not prove their existence in this paper, aside from the theorem of Bishop—Stoll and some standard results on analytic sets (which I take mostly from Łojasiewicz's book [**10**]), the existence of cellular decomposition is the main fact used in my proof of Thereom 5.3, which I give in §5. (I do also use a few standard results on definable sets from [**19, 20**].) I prove Theorem 4.1 in §6.

2. o-minimal structures

In defining o-minimal structures, I follow the book by van den Dries [**19**].

DEFINITION 2.1. An o-minimal structure on $\mathbb{R}$ is a sequence $\mathcal{S} = (\mathcal{S}_n)_{n\in\mathbb{N}}$ of sets such that, for each n:

(1) $\mathcal{S}_n$ is a boolean algebra of subsets of $\mathbb{R}^n$;
(2) $A \in \mathcal{S}_n$ implies that $A \times \mathbb{R}$ and $\mathbb{R} \times A$ are in $\mathcal{S}_{n+1}$;

(3) If $1 \le i < j \le n$, then $\{(x_1, \dots, x_n) \in \mathbb{R}^n : x_i = x_j\} \in \mathcal{S}_n$.
(4) If $\pi : \mathbb{R}^{n+1} \to \mathbb{R}^n$ denotes the projection onto a factor, then $A \in \mathcal{S}_{n+1} \implies \pi(A) \in \mathcal{S}_n$;
(5) For each $r \in \mathbb{R}$, $\{r\} \in \mathcal{S}_1$. Moreover, $\{(x, y) \in \mathbb{R}^2 : x < y\} \in \mathcal{S}_2$;
(6) The only subsets in $\mathcal{S}_1$ are the finite unions of intervals and points.

Call a sequence $\mathcal{S}$ a *structure* if it satisfies all of the hypotheses of Definition 2.1 except possibly the last two [**19**, p.13]. If $X \in \mathcal{S}_n$ for some n, then we say that X is a *definable* subset of $\mathbb{R}^n$ with respect to the structure $\mathcal{S}$. Similarly, if $f : X \to Y$ is a function with $X \subset \mathbb{R}^n$ and $Y \subset \mathbb{R}^m$ for $n, m \in \mathbb{N}$, then we say f is *definable* if its graph (viewed as a subset of $\mathbb{R}^n \times \mathbb{R}^m = \mathbb{R}^{n+m}$) is definable.

It is clear that, if we let $\mathcal{S}_n = \mathcal{P}(\mathbb{R}^n)$, i.e., the power set of $\mathbb{R}^n$, then we get a structure. (But obviously not an o-minimal one.) It is also relatively easy to see that the intersection of structures is a structure. So, given an arbitrary collection $\mathcal{T}_n \subset \mathcal{P}(\mathbb{R}^n)$ (for $n \in \mathbb{N}$), there is a smallest structure $(\mathcal{S}_n)_{n\in\mathbb{N}}$ containing $\mathcal{T}_n$. This is the structure *generated* by the $\mathcal{T}_n$.

If $\{\mathcal{S}_n\}$ and $\{\mathcal{S}'_n\}$ are both structures with $\mathcal{S}'_n \subset \mathcal{S}_n$ for all n, then $\{\mathcal{S}_n\}$ is called an *expansion* of $\{\mathcal{S}'_n\}$. If $\{\mathcal{T}_n\}$ is any collection with $\mathcal{T}_n \subset \mathcal{P}(\mathbb{R}^n)$, the *structure generated by* $\{\mathcal{S}_n \cup \mathcal{T}_n\}_{n\in\mathbb{N}}$ is called the *expansion of* $\{\mathcal{S}_n\}$ *generated by* $\{\mathcal{T}_n\}$.

One classical example of an o-minimal structure on $\mathbb{R}$ is the structure $\mathbb{R}_{\text{alg}}$ consisting of all semi-algebraic sets. (The fact that $\mathbb{R}_{\text{alg}}$ satisfies (4) is the content of the Tarski-Seidenberg theorem.) The example that is most important for the recent applications to Hodge theory mentioned in the introduction is the one called $\mathbb{R}_{\text{an,exp}}$. This is the expansion of $\mathbb{R}_{\text{alg}}$ generated by the graph of the real exponential function $x \mapsto e^x$ and the collection of all graphs of analytic functions on $[0, 1]$. (See [**20**] for references. The o-minimality of the expansion $\mathbb{R}_{\text{exp}}$ of $\mathbb{R}_{\text{alg}}$ generated by the graph of the real exponential function is a celebrated theorem of Wilkie [**21**].)

It is convenient to think about definable sets in a structure in terms of logic as subsets of $\mathbb{R}^n$ defined by the formulas in a language $\mathcal{L}$ interpreted in the field of real numbers. This point of view is explained (a little informally) in [**19**, Chapter 1]. (For a more precise explanation of the model theory point of view, see, for example, Marker's book [**11**]). Here subsets ψ of $\mathbb{R}^n$ are thought of as properties $\psi(x_1, \dots, x_n)$ of n-tuples $(x_1, \dots, x_n)$ of real numbers with $\psi(x_1, \dots, x_n)$ being the property that $(x_1, \dots, x_n) \in \psi$. Suppose $S = \{\psi_i\}$ is a collection of such subsets, with $\psi_i \subset \mathbb{R}^{n_i}$. Then the expansion of $\mathbb{R}_{\text{alg}}$ generated by ψ consists of the subsets of $\mathbb{R}^n$ definable by formulas involving the field operations on $\mathbb{R}$, the real numbers (viewed as constants), the symbols $<$ and $=$, variables $(x_i)_{i=1}^\infty$, and the ψ_i along with the $\forall, \exists$ and the usual logical connectives.

3. Volumes and the Bishop–Stoll Theorem

My main reference for this section is G. Stolzenberg's book [**18**].

Let $X = (X, d_X)$ be a metric space. For $\emptyset \neq S \subset X$, the diameter of S is $\operatorname{diam} S := \sup\{d_X(x, y) : x, y \in S\}$. By convention, write $\operatorname{diam} \emptyset = -\infty$.

Suppose $S \subset X$, and ϵ is a positive real number. An ϵ-covering of S is a countable collection $\{S_i\}_{i=1}^\infty$ of subsets of S of diameter less than ϵ such that $S \subset \bigcup_{i=1}^\infty S_i$. Fix a non-negative real number d and set

$$I(d, \epsilon, S) := \inf \left\{ \sum_{i=1}^{\infty} (\operatorname{diam} S_i)^d : \{S_i\}_{i=1}^\infty \text{ is an } \epsilon\text{-covering of } S \right\}.$$

The *d-Hausdorff measure of* S is

$$\mathcal{H}^d(S) := \frac{1}{2^d} \lim_{\epsilon \to 0^+} I(d, \epsilon, S).$$

If d is a non-negative integer and S is a closed d-dimensional sub-manifold of $\mathbb{R}^n$, then the volume $\operatorname{vol}_d S$ (defined in the usual way with respect to the standard metric on $\mathbb{R}^n$) is given by

$$\operatorname{vol}_d(S) = \frac{\pi^{d/2}}{\Gamma(\frac{d}{2}+1)} \mathcal{H}^d(S). \tag{3.1}$$

As it turns out, we want to refer to volume instead of $\mathcal{H}^d(S)$ in general. So we use the equation (3.1) as a definition to define $\operatorname{vol}_d(S)$ for an arbitrary non-negative real number d. (Note that Federer's normalization for Hausdorff measure in his book [**7**, §2.10.2] differs from that of Stolzenberg. For Federer, the d-dimensional Hausdorff measure $\mathcal{H}^d(S)$ is just what we call $\operatorname{vol}_d(S)$.)

Let's also make the convention that we always regard a subset $S \subset \mathbb{R}^n$ as a metric subspace of $\mathbb{R}^n$ with its standard metric. For a positive real number r, set $B(r) = B_n(r) := \{x \in \mathbb{R}^n : |x| < r\}$. Then, if $S \subset \mathbb{R}^n$, set

$$S(r) := S \cap B(r).$$

We use the Big-O notation: if f, g are two real valued functions defined on an interval of the form (a, ∞), then we write $f = O(g)$ if there exists a constant C and a real number $b > a$ such that

$$x > b \Rightarrow |f(x)| \leq Cg(x).$$

THEOREM 3.1 (Stoll). *Suppose that Z is a closed analytic subset of $\mathbb{C}^n$ of pure dimension d. If $\operatorname{vol}_{2d} Z(r) = O(r^{2d})$, then Z is algebraic.*

See [**18**, p. 2, Theorem D] for the statement. A proof is given in [**18**, Chapter IV]. According to Cornalba and Griffiths [**6**, E. 4.2], the converse also holds. (This also follows directly from the main result of this note, Theorem 4.1 below.)

4. Volumes of Definable Sets

My main goal in this note is to prove the Peterzil–Starchenko GAGA theorem (Theorem 5.3 below) using Theorem 3.1 and a general fact about definable sets and Hausdorff measures. To explain this general fact, let me first explain cells. To do this, fix an o-minimal expansion $\mathbb{R}_{\text{alg},*}$ of $\mathbb{R}_{\text{alg}}$.

4.1. Cells. These are certain special definable subsets (with respect to $\mathbb{R}_{\text{alg},*}$) of $\mathbb{R}^n$ defined inductively. See page 50 of [**19**] for a complete definition, but, roughly speaking, cells in $\mathbb{R}^n$ are defined inductively with respect to n as either

(a) graphs of continuous definable functions f on cells in $\mathbb{R}^{n-1}$ or,
(b) nonempty open regions in between graphs of continuous definable functions in $\mathbb{R}^{n-1}$.

There is a dimension function d defined inductively on the set of all cells by setting $d(S) = d(T)$ if $S \subset \mathbb{R}^n$ is constructed inductively from $T \subset \mathbb{R}^{n-1}$ via procedure (a) and setting $d(S) = d(T) + 1$ if it is constructed via procedure (b). Moreover, as a consequence of the cell decomposition theorem [**19**, 2.11 on p. 52], every definable subset of $\mathbb{R}^n$ can be written as a finite disjoint union of cells. (This is one of the most crucial properties of definable sets in o-minimal structures.) If X is then any definable subset of $\mathbb{R}^n$, van den Dries defines $\dim X$ to be the maximum of the dimensions $d(S)$ as S ranges over all cells contained in X [**19**, p. 63]. Since, by [**19**, p. 64], $\dim(X \cup Y) = \max(\dim X, \dim Y)$, $\dim X$ is also the maximum of the dimensions $d(S)$ of the cells S appearing in a decomposition of X into disjoint cells.

Now, I am ready to state the main volume estimate of this paper. As mentioned in the introduction, this theorem is a special case of Proposition 3.1 of [**13**]. Moreover, the paper [**9**] by Kurdyka and Raby proves an equivalent result in the language of subanalytic subsets.

THEOREM 4.1. *Suppose $S \subset \mathbb{R}^n$ is a set which is definable in an o-minimal expansion of $\mathbb{R}_{\mathrm{alg}}$. Set $d = \dim S$. Then*

$$\mathrm{vol}_d\, S(r) = O(r^d).$$

5. Peterzil–Starchenko

Before proving Theorem 4.1, I want to use it to prove the Peterzil–Starchenko GAGA theorem (Theorem 5.3) using Theorems 4.1 and Theorem 3.1. This would be very straightforward were it not for the hypothesis in Theorem 3.1 that the analytic subset Z is pure dimensional. The most direct way to get around this nuisance seems to be to prove that a definable analytic subset is a finite union of irreducible components, each of which is definable, analytic and pure dimensional. This is what I do in Lemma 5.2 below.

First, it will be help to recall some of the notation and results from Łojasiwicz's book on complex analytic geometry, starting from [**10**, p. 208]. Suppose A is a locally analytic subset of a n-dimensional complex manifold M. This means that every point $x \in A$ has an open neighborhood U such that $A \cap U$ is a (closed) analytic subset of U [**10**, p. 156]. Write A^*, A^0 and $A^{(k)}$ for the set of singular points of A, the set of regular points of A , and the set of regular points of dimension k of A respectively. If A is an analytic subset of M, then A is said to be *simple* or *irreducible* if it cannot be written as the union of two proper analytic subsets. In fact, A being irreducible is equivalent to A^0 being connected [**10**, p. 215, Proposition 2]. (Łojasiewicz uses the word *simple* interchangeably with the word *irreducible*, but I prefer to just use the word *irreducible*.)

According to [**10**, p. 217], a *decomposition* of an analytic subset $A \subseteq M$ into irreducible components is an irredundant representation $A = \bigcup A_i$ of A as a locally finite union of irreducible analytic subsets A_i. Here *irredundant* means that, for each $i \neq j$, $A_i \not\subset A_j$. Given this, we then have the following unique decomposition theorem [**10**, p. 217, Theorem 4].

THEOREM 5.1. *Every analytic subset $A \subseteq M$ has a unique decomposition into irreducible components A_i. If we set $U_i := A_i \setminus A^*$, then $U_i = A_i^0 \setminus A^*$ and U_i is open and dense in both A_i^0 and in A_i. Moreover, the U_i are exactly the connected components of A^0.*

Note that irreducible analytic sets are necessary pure dimensional [**10**, p.216, Corollary 1].

LEMMA 5.2. *Suppose A is a closed, complex analytic subset of $\mathbb{C}^n$, which is definable with respect respect to an o-minimal expansion of $\mathbb{R}_{\mathrm{alg}}$. Then the decomposition in Theorem 5.1 is finite, and the U_i as well as the A_i are definable.*

PROOF. To prove this, we use some results about the regular locus of A viewed as a definable subset of $\mathbb{R}^{2n} = \mathbb{C}^n$. In other words, we use some results about A as a definable set forgetting its analytic structure.

Following the notation of [**20**], write $\operatorname{Reg}_k^p A$ for the set of points of A where A is locally a C^p submanifold of $\mathbb{C}^n$ of dimension k. By [**20**, p. 519, B.9], for each integer k and every positive integer p, $\operatorname{Reg}_k^p A$ is a definable. On the other hand, from a result from the beginning of Milnor's book [**12**, p.13], it follows that the locus $A^{(k)}$ defined above is precisely equal to $\operatorname{Reg}_{2k}^1 A$. In other words, the locus of regular points of dimension k of A viewed as a complex analytic space is exactly equal to the locus of points where A is a $2k$-dimensional C^1 submanifold of $\mathbb{C}^n$. (And, obviously, this implies that, in fact, $\operatorname{Reg}_k^1 A = \operatorname{Reg}_k^p A$ for $p \geq 1$.) See [**15**, p. 52] for another explanation of this fact.

It follows that each subset $A^{(k)}$ is a definable submanifold of $\mathbb{C}^n$. But then, by [**20**, p. 500, 1.10-1.14], $A^{(k)}$ has finitely many connected components, each of which is definable. Since the U_i are the connected components of A^0, it follows that there are finitely many U_i and that each U_i is definable. Therefore, by [**19**, Lemma 3.4], the closure A_i of U_i is definable as well. □

THEOREM 5.3 (Peterzil–Starchenko). *Let A be a closed, complex analytic subset of $\mathbb{C}^n$, which is definable with respect to an o-minimal expansion of $\mathbb{R}_{\text{alg}}$. Then A is an algebraic subset of $\mathbb{C}^n$.*

PROOF OF THEOREM 5.3 USING THEOREM 4.1. Take a closed complex analytic subset $A \subseteq \mathbb{C}^n$, and assume A is definable. By Lemma 5.2, we can write $A = \bigcup_{i=1}^m A_i$ with A_i definable, analytic, and irreducible. From this, we see that it suffices to prove the Theorem in the case that A is irreducible.

So assume that A is irreducible. Then, as noted before the statement of Lemma 5.2, A is of pure (complex) dimension d for some integer d. Then d is also the complex dimension of A^0. From this, it is easy to see that $2d$ is the dimension of A as a definable subset of $\mathbb{C}^n$ (in the sense of [**19**, Definition 4.1.1]). Therefore, by Theorem 4.1, we have $\operatorname{vol}_{2d} A(r) = O(r^{2d})$. So, by Theorem 3.1, A is algebraic. □

In the next section, I prove Theorem 4.1. The elementary proof mainly relies on the Gauss map, change of variables and the existence of cellular decompositions.

6. Proof of the Volume Estimate

In what follows it will be convenient to note that, since linear transformations between finite dimensional real vector spaces are definable in any expansion of $\mathbb{R}_{\text{alg}}$, any finite dimensional real vector space V comes equipped with a canonical definable structure (via any linear isomorphism to $\mathbb{R}^{\dim V}$).

If X is a definable subset of $\mathbb{R}^n$ of dimension d, we say that *Theorem* 4.1 *holds for* X if $\operatorname{vol}_d X(r) = O(r^d)$. Note that, if Theorem 4.1 holds for X, then, for $d' > d$, we have $\operatorname{vol}_{d'} X(r) = 0$ for all r. (See [**7**, §2.10.2].)

Write $\operatorname{Gr}(d, n)$ for the Grassmannian of real d-dimensional planes through the origin in $\mathbb{R}^n$. The set $\operatorname{Gr}(d, n)$ has a natural structure of a definable C^∞-manifold. In the language of [**19**, Chapter 10], $\operatorname{Gr}(d, n)$ is a definable space, which is also (compatibly) a compact C^∞-manifold. (See also [**8**] for a precise definition.) As it is also a regular space, [**19**, Theorem 10.1.8] implies that it is isomorphic (as a definable space) to an affine definable space, i.e., a definable subset of $\mathbb{R}^n$. On the other hand, the method used in [**19**, Example 10.1.4] to show that $\mathbb{P}^n(\mathbb{R})$ is affine, can be imitated to realize $\mathrm{Gr}(d,n)$ as a closed definable C^∞-submanifold of $\mathbb{R}^N$ for some suitable N.

To be explicit about this last point, endow $\mathbb{R}^n$ with the usual dot product. For each $L \in \mathrm{Gr}(d,n)$, choose an ordered basis $\tilde{L} = (\ell_1, \ldots, \ell_d)$, and set $\wedge^d \tilde{L} := \ell_1 \wedge \cdots \wedge \ell_d$. Then let $\phi(L)$ denote the point

$$\phi(L) := \frac{1}{\|\wedge^d \tilde{L}\|^2} (\wedge^d \tilde{L}) \otimes (\wedge^d \tilde{L}) \in \mathrm{Sym}^2(\wedge^d \mathbb{R}^n).$$

The resulting map $\phi : \mathrm{Gr}(d,n) \to \mathrm{Sym}^2(\wedge^d \mathbb{R}^n)$ is a well-defined and definable smooth morphism, embedding $\mathrm{Gr}(d,n)$ as a closed, definable, smooth submanifold of $\mathrm{Sym}^2(\wedge^d \mathbb{R}^n)$. So we can identify $\mathrm{Gr}(d,n)$ with the image of ϕ.

We can also view $\mathrm{Gr}(d,n)$ as a quotient of the orthogonal group $\mathbf{O}(n)$ in the usual way. Note that the orthogonal group $\mathbf{O}(n)$ (of real $n \times n$-matrices which are orthogonal with respect to the standard inner product) is a closed definable and smooth submanifold of $\mathbb{R}^{n^2}$. Moreover, it is a group in the category of definable spaces. It acts definably, properly and transitively on the space $\mathrm{Gr}(d,n)$. The stabilizer of $L \in \mathrm{Gr}(d,n)$ is the definable, closed subgroup $\mathbf{O}(L) \times \mathbf{O}(L^\perp)$, which is definably isomorphic to $\mathbf{O}(d) \times \mathbf{O}(n-d)$. From this, it is not hard to see that, in the language of [**19**, p. 162], $\mathrm{Gr}(d,n)$ is a definably proper quotient of $\mathbf{O}(n)$, and, in fact, $\mathrm{Gr}(d,n)$ is definably isomorphic to the quotient $\mathbf{O}(n)/(\mathbf{O}(d) \times \mathbf{O}(n-d))$.

If $L \in \mathrm{Gr}(d,n)$ is a d-dimensional linear subspace, we write $\pi_L : \mathbb{R}^n \to L$ for the orthogonal projection onto the subspace L. The map $z \mapsto (\pi_L(z), \pi_{L^\perp}(z))$ is then an isometric (and, thus, volume-preserving) definable isomorphism from $\mathbb{R}^n$ to $L \times L^\perp$.

Definition 6.1. Suppose $L \in \mathrm{Gr}(d,n)$ is a subspace of $\mathbb{R}^n$. We say that a d-dimensional definable subset M of $\mathbb{R}^n$ is a *good L graph* if there exists

(1) an open d-dimensional cell C in L,
(2) a differentiable definable function $f : C \to L^\perp$,
(3) a real constant K with $\|f'(x)\| \le K$ for all $x \in C$,

such that

(4) $M = \{z \in \mathbb{R}^n : \pi_L(z) \in C, \pi_{L^\perp}(z) = f(\pi_L(z))\}$.

We say that M is a *good graph* if it is a good graph relative to some d-dimensional subspace L of $\mathbb{R}^n$. We say that M is an *L graph* if it satisfies conditions (1), (2) and (4) but not necessarily (3).

Remarks 6.2. (a) In case $L = \{(z_1, \ldots, z_n) \in \mathbb{R}^n : z_i = 0 \text{ for } i > d\}$, a good L graph is just the graph of a definable, differentiable function $f : C \to \mathbb{R}^{n-d}$ on a cell C such that an inequality of the form $\|f'(x)\| \le K$ is satisfied globally on C.

(b) If C is convex, then Definition 6.1 (iii) implies that the function f in the definition of a good L graph is Lipschitz.

Lemma 6.3. *Theorem* 4.1 *holds for any good graph.*

Proof. Since the orthogonal group $\mathbf{O}(n)$ acts transitively on $\mathrm{Gr}(d,n)$ and preserves volumes and the property of being a good graph, we can assume $L = \{(x_1, \ldots, x_n) \in \mathbb{R}^n : x_i = 0 \text{ for } i > d\}$, and that M is the graph of $f : C \to \mathbb{R}^{n-d}$, where C is an open cell in L. Write e_i for the tangent vector $\partial/\partial x_i$, and write $\Phi : C \to \mathbb{R}^n$ for the map $x \mapsto (x, f(x))$. Let $A(x) = f'(x)$ denote the derivative of

f at x. For $z = (x, f(x)) \in M$, the tangent space $T_z M$ is the d-dimensional space generated by the vectors $v_i = e_i + A(x)e_i \in \mathbb{R}^n$. For $r \in \mathbb{R}$, $\pi_L(M(r)) \subset C(r)$. Therefore, $\mathrm{vol}_d\, M(r) \leq \int_{C(r)} \sqrt{\det g_{ij}(x)}\, dx$ where $g_{ij}(x)$ is the matrix $v_i \cdot v_j = \delta_{ij} + e_i \cdot A(x)^* A(x) e_j$. (See [**7**, §3.2.46] for the relevant formula computing the volume in terms of g_{ij}.) Since M is a good graph, $\|A(x)\|$ and thus $g_{ij}(x)$ is uniformly bounded on C. From these considerations, the lemma follows easily. □

LEMMA 6.4. *Every $L \in \mathrm{Gr}(d, n)$ has an open neighborhood $U_L \subseteq \mathrm{Gr}(d, n)$ such that the following conditions hold:*

(1) *$\pi_L(L') = L$ for all $L' \in U_L$;*
(2) *if M is an L-graph with $T_z M \in U_L$ for all $z \in M$, then M is a good L graph.*

PROOF. Write $b_1 = (1, 0, \ldots, 0)$, $b_2 = (0, 1, \ldots, 0), \ldots, b_n = (0, \ldots, 0, 1)$ for the standard basis vectors in $\mathbb{R}^n$. As in Lemma 6.3 we can assume that $L = \langle b_1, \ldots, b_d \rangle$, and that $f : C \to \mathbb{R}^{n-d}$, where C is an open cell in L. Write $\mathrm{Hom}(\mathbb{R}^d, \mathbb{R}^{n-d})$ for the vector space of linear maps from the vector space $\mathbb{R}^d$ to the vector space $\mathbb{R}^{n-d}$, and write $\Phi : \mathrm{Hom}(\mathbb{R}^d, \mathbb{R}^{n-d}) \to \mathrm{Gr}(d, n)$ for the map sending a transformation T to the subspace $\langle e_1 + T(e_1), \ldots, e_d + T(e_d) \rangle$ spanned by the vectors $e_i + Te_i$. The map Φ defines a homeomorphism from $\mathrm{Hom}(\mathbb{R}^d, \mathbb{R}^{n-d})$ onto the open neighborhood V of L consisting of all subspaces L' such that $\pi_L(L') = L$. Now take $U_L = \Phi(W)$ where W is any bounded open neighborhood of 0 in $\mathrm{Hom}(\mathbb{R}^d, \mathbb{R}^{n-d})$. If M is an L-graph with $T_z M \in U_L$, then $f'(z) \in W$ for all $z \in C$. So $\|f'(z)\|$ is uniformly bounded on C. Therefore, M is a good L graph. Thus, conditions (1) and (2) both hold. □

For each definable set X, we let $\mathrm{Reg}^2 X$ denote the locus in X consisting of all points $x \in X$ such that there is an open neighborhood U of x in $\mathbb{R}^n$ such that $U \cap X$ is a C^2-manifold. This is a definable subset of X, and the complement $X \setminus \mathrm{Reg}^2 X$ has dimension strictly less than the dimension of X. (This follows from the Cell Decomposition Theorem of [**20**, §4.2].)

PROPOSITION 6.5. *Suppose X is a definable subset of $\mathbb{R}^n$ of dimension d. Then X is a finite union of d'-dimensional good graphs with $d' \leq d$.*

PROOF. We can write X as a finite union of cells of dimension $\leq d$. Moreover, using the remarks above, we can assume that each of these cells is a C^2 manifold. Working by induction on d, we can then assume that X consists of only one cell, which is a C^2 manifold of dimension d.

We then get a definable continuous map $G : X \to \mathrm{Gr}(d, n)$ given by $z \mapsto T_z X$. (I call it G for Gauss, as it is sometimes known as the Gauss map.) Since $\mathrm{Gr}(d, n)$ is compact, we can find a finite cover U_i of $\mathrm{Gr}(d, n)$ such that each U_i is an open neighborhood of the form U_L as in Proposition 6.4. Setting $X_i = G^{-1}(U_i)$ it suffices to show that each X_i is a finite union of d'-dimensional good graphs with $d' \leq d$. So we can replace X with X_i and assume that $G(X)$ is contained in one U_i. Then, using the transitivity of $\mathbf{O}(n)$ on $\mathrm{Gr}(d, n)$, we can assume that $U_i = U_{L_0}$ where $L_0 = \langle b_0, \ldots, b_d \rangle$ is the span of the first d standard basis vectors of $\mathbb{R}^n$.

Since X is a both a d-dimensional cell and a C^2 manifold, and since $\pi_{L_0}(T_z X) = L_0$ for all $z \in X$, it follows that X is an L_0 graph. Then, by Lemma 6.4, it follows that X is good. So the result is proved. □

PROOF OF THEOREM 4.1. Suppose S is a definable set of dimension d. By Proposition 6.5, S is a finite union of d'-dimensional good graphs with $d' \leq d$. Then by Lemma 6.3, Theorem 4.1 holds for any good graph. So, since a finite sum of $O(r^{d'})$ functions for $d' \leq d$ is in $O(r^d)$, the result follows. □

Acknowledgments

I thank my colleague Chris Laskowski for several enlightening conversations about model theory in general and o-minimal structures in particular. I am also happy to thank Najmuddin Fakhruddin, Priyankur Chaudhuri, Eoin Mackall, Swarnava Mukhopadhyay as well the anonymous referee for suggestions and corrections they pointed out after reading previous versions of this paper. I am particularly grateful to the referee for pointing out a mistake in my original proof of Theorem 5.3. (Essentially, I was applying Theorem 3.1 in a situation where the analytic set Z is not pure dimensional without checking first that this can be justified.) Finally, I am extremely grateful to Xuan Viet Nhan Nguyen, Tobias Kaiser and Georges Compte for alerting me to the history discussed above in §1.1.

References

[1] B. Bakker, B. Klingler, and J. Tsimerman, *Tame topology of arithmetic quotients and algebraicity of Hodge loci*, J. Amer. Math. Soc. **33** (2020), no. 4, 917–939, DOI 10.1090/jams/952. MR4155216

[2] Benjamin Bakker, Yohan Brunebarbe, and Jacob Tsimerman, *o-minimal GAGA and a conjecture of Griffiths*, Invent. Math. **232** (2023), no. 1, 163—-228, DOI 10.1007/s00222-022-01166-1. MR4557401

[3] Errett Bishop, *Conditions for the analyticity of certain sets*, Michigan Math. J. **11** (1964), 289–304. MR168801

[4] Eduardo Cattani, Pierre Deligne, and Aroldo Kaplan, *On the locus of Hodge classes*, J. Amer. Math. Soc. **8** (1995), no. 2, 483–506, DOI 10.2307/2152824. MR1273413

[5] E. M. Chirka, *Complex analytic sets*, Mathematics and its Applications (Soviet Series), vol. 46, Kluwer Academic Publishers Group, Dordrecht, 1989. Translated from the Russian by R. A. M. Hoksbergen, DOI 10.1007/978-94-009-2366-9. MR1111477

[6] Maurizio Cornalba and Phillip Griffiths, *Analytic cycles and vector bundles on non-compact algebraic varieties*, Invent. Math. **28** (1975), 1–106, DOI 10.1007/BF01389905. MR367263

[7] Herbert Federer, *Geometric measure theory*, Die Grundlehren der mathematischen Wissenschaften, Band 153, Springer-Verlag New York, Inc., New York, 1969. MR0257325

[8] Andreas Fischer, *Smooth functions in o-minimal structures*, Adv. Math. **218** (2008), no. 2, 496–514, DOI 10.1016/j.aim.2008.01.002. MR2407944

[9] K. Kurdyka and G. Raby, *Densité des ensembles sous-analytiques* (French, with English summary), Ann. Inst. Fourier (Grenoble) **39** (1989), no. 3, 753–771. MR1030848

[10] Stanisław Łojasiewicz, *Introduction to complex analytic geometry*, Birkhäuser Verlag, Basel, 1991. Translated from the Polish by Maciej Klimek, DOI 10.1007/978-3-0348-7617-9. MR1131081

[11] David Marker, *Model theory*, Graduate Texts in Mathematics, vol. 217, Springer-Verlag, New York, 2002. An introduction. MR1924282

[12] John Milnor, *Singular points of complex hypersurfaces*, Annals of Mathematics Studies, No. 61, Princeton University Press, Princeton, N.J.; University of Tokyo Press, Tokyo, 1968. MR0239612

[13] Nhan Nguyen and Guillaume Valette, *Whitney stratifications and the continuity of local Lipschitz-Killing curvatures* (English, with English and French summaries), Ann. Inst. Fourier (Grenoble) **68** (2018), no. 5, 2253–2276. MR3893769

[14] Ya'acov Peterzil and Sergei Starchenko, *Complex analytic geometry in a nonstandard setting*, Model theory with applications to algebra and analysis. Vol. 1, London Math. Soc. Lecture Note Ser., vol. 349, Cambridge Univ. Press, Cambridge, 2008, pp. 117–165, DOI 10.1017/CBO9780511735226.008. MR2441378

[15] Ya'acov Peterzil and Sergei Starchenko, *Complex analytic geometry and analytic-geometric categories*, J. Reine Angew. Math. **626** (2009), 39–74, DOI 10.1515/CRELLE.2009.002. MR2492989

[16] Bernard Shiffman, *On the removal of singularities of analytic sets*, Michigan Math. J. **15** (1968), 111–120. MR224865

[17] Wilhelm Stoll, *The growth of the area of a transcendental analytic set. I, II*, Math. Ann. **156** (1964), 144–170, DOI 10.1007/BF01359928. MR166393

[18] Gabriel Stolzenberg, *Volumes, limits, and extensions of analytic varieties*, Lecture Notes in Mathematics, No. 19, Springer-Verlag, Berlin-New York, 1966. MR0206337

[19] Lou van den Dries, *Tame topology and o-minimal structures*, London Mathematical Society Lecture Note Series, vol. 248, Cambridge University Press, Cambridge, 1998, DOI 10.1017/CBO9780511525919. MR1633348

[20] Lou van den Dries and Chris Miller, *Geometric categories and o-minimal structures*, Duke Math. J. **84** (1996), no. 2, 497–540, DOI 10.1215/S0012-7094-96-08416-1. MR1404337

[21] A. J. Wilkie, *Model completeness results for expansions of the ordered field of real numbers by restricted Pfaffian functions and the exponential function*, J. Amer. Math. Soc. **9** (1996), no. 4, 1051–1094, DOI 10.1090/S0894-0347-96-00216-0. MR1398816

Department of Mathematics, University of Maryland, College Park, Maryland
Email address: pbrosnan@umd.edu

Contemporary Mathematics
Volume **790**, 2023
https://doi.org/10.1090/conm/790/15857

Hessenberg varieties and Poisson slices

Peter Crooks and Markus Röser

ABSTRACT. This expository article considers a circle of Lie-theoretic ideas involving Hessenberg varieties, Poisson geometry, and wonderful compactifications. In more detail, one may associate a symplectic Hamiltonian G-variety $\mu : G\times\mathcal{S} \longrightarrow \mathfrak{g}$ to each complex semisimple Lie algebra $\mathfrak{g}$ with adjoint group G and fixed Kostant section $\mathcal{S} \subseteq \mathfrak{g}$. This variety is one of Bielawski's hyperkähler slices, and it is central to Moore and Tachikawa's work on topological quantum field theories. It also bears a close relation to two log symplectic Hamiltonian G-varieties $\overline{\mu}_{\mathcal{S}} : \overline{G\times\mathcal{S}} \longrightarrow \mathfrak{g}$ and $\nu : \mathrm{Hess} \longrightarrow \mathfrak{g}$. The former is a Poisson transversal in the log cotangent bundle of the wonderful compactification $\overline{G}$, while the latter is the standard family of Hessenberg varieties. Each of $\overline{\mu}$ and ν is known to be a fibrewise compactification of μ.

We exploit the theory of Poisson slices to relate the fibrewise compactifications mentioned above. Our work is shown to be compatible with a Poisson isomorphism obtained by Bălibanu.

CONTENTS

1. Introduction

1.1. Context. Let $\mathfrak{g}$ be a complex semisimple Lie algebra with adjoint group G and fixed Kostant section $\mathcal{S} \subseteq \mathfrak{g}$. These data determine a symplectic Hamiltonian G-variety $\mu : G\times\mathcal{S} \longrightarrow \mathfrak{g}$ that has received considerable attention in the literature; it is perhaps the most fundamental of Bielawski's *hyperkähler slices* [**8, 9, 18**], and it features prominently in Moore and Tachikawa's work on topological quantum field theories [**32**]. Some more recent work is concerned with the moment map $\mu : G \times \mathcal{S} \longrightarrow \mathfrak{g}$ and its geometric properties [**2, 15, 16**]. One such property is the fact that μ admits two fibrewise compactifications; the first is a well-studied family of Hessenberg varieties, while the other is constructed via the wonderful compactification $\overline{G}$ of G. The following discussion gives some context for the two compactifications.

2020 *Mathematics Subject Classification.* Primary 14L30; Secondary 53D20, 14M17.

Hessenberg varieties arise as a natural generalization of Grothendieck–Springer fibres, and their study is central to modern research in algebraic geometry [**3, 13, 19, 33, 38**], combinatorics [**1, 4, 27, 37, 39**], representation theory [**5–7**], and symplectic geometry [**2, 30**]. One begins by fixing a Borel subgroup $B \subseteq G$ with Lie algebra $\mathfrak{b} \subseteq \mathfrak{g}$. A *Hessenberg subspace* is then defined to be a B-invariant vector subspace $H \subseteq \mathfrak{g}$ that contains $\mathfrak{b}$. Each Hessenberg subspace $H \subseteq \mathfrak{g}$ determines a G-equivariant vector bundle $G \times_B H \longrightarrow G/B$, and the total space of this bundle is Poisson. The G-action on $G \times_B H$ is Hamiltonian and admits an explicit moment map $\nu_H : G \times_B H \longrightarrow \mathfrak{g}$. One calls

$$\mathrm{Hess}(x, H) := \nu_H^{-1}(x)$$

the *Hessenberg variety* associated to H and $x \in \mathfrak{g}$, and regards ν_H as the family of all Hessenberg varieties associated to H.

The so-called *standard Hessenberg subspace* is the annihilator $\mathfrak{m}$ of $[\mathfrak{u}, \mathfrak{u}]$ under the Killing form, where $\mathfrak{u}$ is the nilradical of $\mathfrak{b}$. The resulting family $\nu := \nu_{\mathfrak{m}} : G \times_B \mathfrak{m} \longrightarrow \mathfrak{g}$ is of particular importance. One reason is that the fibres of ν appear in interesting contexts; $\mathrm{Hess}(x, \mathfrak{m})$ is isomorphic to the Peterson variety if $x \in \mathfrak{g}$ is regular and nilpotent, while $\mathrm{Hess}(x, \mathfrak{m})$ is a well-studied smooth projective toric variety if x is regular and semisimple. A second reason is that $G \times_B \mathfrak{m}$ enjoys a rich Poisson geometry; Bălibanu has shown this variety to be log symplectic [**5**], while Abe and the first author have identified $G \times \mathcal{S}$ as the unique open dense symplectic leaf in $G \times_B \mathfrak{m}$ [**2**]. It is in this context that one realizes $\nu : G \times_B \mathfrak{m} \longrightarrow \mathfrak{g}$ as a fibrewise compactification of $\mu : G \times \mathcal{S} \longrightarrow \mathfrak{g}$.

The connection between $\mu : G \times \mathcal{S} \longrightarrow \mathfrak{g}$ and $\overline{G}$ is realized via $T^*\overline{G}(\log D)$, the *log cotangent bundle* of $\overline{G}$. This variety is log symplectic, and it contains a distinguished log symplectic subvariety $\overline{G \times \mathcal{S}} \subseteq T^*\overline{G}(\log D)$. The subvariety $\overline{G \times \mathcal{S}}$ is a so-called *Poisson slice* [**17**] carrying a Hamiltonian G-action and moment map $\overline{\mu} : \overline{G \times \mathcal{S}} \longrightarrow \mathfrak{g}$, and it contains $G \times \mathcal{S}$ as its unique open dense symplectic leaf. The moment map $\overline{\mu} : \overline{G \times \mathcal{S}} \longrightarrow \mathfrak{g}$ is thereby a fibrewise compactification of $\mu : G \times \mathcal{S} \longrightarrow \mathfrak{g}$.

It is natural to seek a relationship between the log symplectic fibrewise compactifications $\nu : G \times_B \mathfrak{m} \longrightarrow \mathfrak{g}$ and $\overline{\mu} : \overline{G \times \mathcal{S}} \longrightarrow \mathfrak{g}$ of $\mu : G \times \mathcal{S} \longrightarrow \mathfrak{g}$. We are thereby drawn to Bălibanu's recent article [**5**], which studies the *universal centralizer* $\mathcal{Z}_{\mathfrak{g}} \longrightarrow \mathcal{S}$. Bălibanu takes an appropriate Whittaker reduction of $T^*\overline{G}(\log(D))$ and obtains a log symplectic fibrewise compactification $\overline{\mathcal{Z}_{\mathfrak{g}}} \longrightarrow \mathcal{S}$ of the universal centralizer. She subsequently shows $\nu^{-1}(\mathcal{S})$ to be Poisson, and to be isomorphic to $\overline{\mathcal{Z}_{\mathfrak{g}}}$ as a Poisson variety over $\mathcal{S}$.

1.2. Summary of results. We use the theory of Poisson slices [**17**] to construct a canonical bimeromorphism $\overline{G \times \mathcal{S}} \cong G \times_B \mathfrak{m}$ and explain its relation to Bălibanu's Poisson isomorphism. To this end, fix a Borel subgroup B with Lie algebra $\mathfrak{b}$. Let us also fix a principal $\mathfrak{sl}_2$-triple $\tau = (\xi, h, \eta) \in \mathfrak{g}^{\oplus 3}$ with $\eta \in \mathfrak{b}$. One then has an associated Slodowy slice (a.k.a. Kostant section)

$$\mathcal{S} := \xi + \mathfrak{g}_\eta \subseteq \mathfrak{g}.$$

The following is our main result.

THEOREM 1.1. *Subject to a gluing condition described in Remark* 4.11, *the following statements hold.*

(i) *There exists a Hamiltonian G-variety isomorphism*

$$\overline{G\times\mathcal{S}}^{\circ}\xrightarrow{\cong}(G\times_B\mathfrak{m})^{\circ},$$

where $\overline{G\times\mathcal{S}}^{\circ}\subseteq\overline{G\times\mathcal{S}}$ *and* $(G\times_B\mathfrak{m})^{\circ}\subseteq G\times_B\mathfrak{m}$ *are explicit G-invariant open dense subvarieties with complements of codimension at least two.*

(ii) *The isomorphism in* (i) *restricts to a variety isomorphism*

$$\overline{G\times\mathcal{S}}^{\circ}\cap\overline{\mathcal{Z}_{\mathfrak{g}}}\xrightarrow{\cong}(G\times_B\mathfrak{m})^{\circ}\cap\nu^{-1}(\mathcal{S}).$$

This restricted isomorphism agrees with the restriction of Bălibanu's Poisson isomorphism to $\overline{G\times\mathcal{S}}^{\circ}\cap\overline{\mathcal{Z}_{\mathfrak{g}}}$.

(iii) *The isomorphism in* (i) *extends to a G-equivariant bimeromorphism*

$$\overline{G\times\mathcal{S}}\xrightarrow{\cong}G\times_B\mathfrak{m}$$

for which

$$\begin{array}{ccc}\overline{G\times\mathcal{S}} & \xrightarrow{\quad\cong\quad} & G\times_B\mathfrak{m}\\ & {\scriptstyle\overline{\mu}_{\mathcal{S}}}\searrow\quad\swarrow{\scriptstyle\nu} & \\ & \mathfrak{g} & \end{array}$$

commutes.

(iv) *The bimeromorphism in* (iii) *is a biholomorphism if* $\mathfrak{g}=\mathfrak{sl}_2$.

This theorem leads to the following conjecture.

Conjecture 1.2. The bimeromorphism in Theorem 1.1(iii) is a biholomorphism.

We show this conjecture to have the following implications for Hessenberg varieties.

Corollary 1.3. *Assume that Conjecture* 1.2 *is true, and let* $\overline{\rho}:\overline{G\times\mathcal{S}}\longrightarrow G\times_B\mathfrak{m}$ *be the biholomorphism from Theorem* 1.1(iii). *Let* $\pi:T^*\overline{G}(\log(D))\longrightarrow\overline{G}$ *be the bundle projection map. If* $x\in\mathfrak{g}$, *then the composite map*

$$\mathrm{Hess}(x,\mathfrak{m})\xrightarrow{\overline{\rho}^{-1}}\overline{\mu}_{\mathcal{S}}^{-1}(x)\xrightarrow{\pi}\overline{G}$$

is a G_x-equivariant closed embedding of algebraic varieties.

More succinctly, the truth of Conjecture 1.2 would imply that every Hessenberg variety in the family $\nu:G\times_B\mathfrak{m}\longrightarrow\mathfrak{g}$ admits an equivariant closed embedding into $\overline{G}$.

Remark 1.4. A recently revised version of Bălibanu's preprint [**5**] includes a proof of Conjecture 1.2. This revision and proof appeared after our manuscript had been posted on arXiv.

1.3. Organization. Section 2 assembles some of the Lie-theoretic and Poisson-geometric facts, conventions, and notation underlying this paper. The heart of our paper begins in Section 3, which is principally concerned with the log symplectic variety $\overline{G\times\mathcal{S}}$ and its properties. Section 4 expands our discussion to include implications for the family $\nu:G\times_B\mathfrak{m}\longrightarrow\mathfrak{g}$. This section contains the proofs of Theorem 1.1 and Corollary 1.3, as well as the requisite machinery. A brief list of recurring notation appears after Section 4.

2. Some preliminaries

This section gathers some of the notation, conventions, and standard facts used throughout our paper. Our principal objective is to outline the Lie-theoretic constructions relevant to later sections.

2.1. Fundamental conventions. This paper works exclusively over $\mathbb{C}$. We understand *group action* as meaning *left group action*. The dimension of an algebraic variety is the supremum of the dimensions of its irreducible components. We use the term *smooth variety* in reference to a pure-dimensional algebraic variety X satisfying $\dim(T_xX) = \dim X$ for all $x \in X$.

2.2. Lie-theoretic conventions. Let $\mathfrak{g}$ be a finite-dimensional, rank-ℓ, semisimple Lie algebra with adjoint group G. Write

$$\mathrm{Ad} : G \longrightarrow \mathrm{GL}(\mathfrak{g}), \quad g \mapsto \mathrm{Ad}_g, \quad g \in G$$

for the adjoint representation of G on $\mathfrak{g}$, and

$$\mathrm{ad} : \mathfrak{g} \longrightarrow \mathfrak{gl}(\mathfrak{g}), \quad x \mapsto \mathrm{ad}_x, \quad x \in \mathfrak{g}$$

for the adjoint representation of $\mathfrak{g}$ on itself. Each $x \in \mathfrak{g}$ admits a G-stabilizer

$$G_x := \{g \in G : \mathrm{Ad}_g(x) = x\}$$

and a $\mathfrak{g}$-centralizer

$$\mathfrak{g}_x := \ker(\mathrm{ad}_x) = \{y \in \mathfrak{g} : [x, y] = 0\}.$$

One also has the G-invariant open dense subvariety

$$\mathfrak{g}^{\mathrm{r}} := \{x \in \mathfrak{g} : \dim(\mathfrak{g}_x) = \ell\}$$

of all regular elements in $\mathfrak{g}$.

One calls $x \in \mathfrak{g}$ *semisimple* (resp. *nilpotent*) if $\mathrm{ad}_x \in \mathfrak{gl}(\mathfrak{g})$ is diagonalizable (resp. nilpotent) as a vector space endomorphism. Let $\mathfrak{g}^{\mathrm{rs}}$ denote the G-invariant open dense subvariety of regular semisimple elements in $\mathfrak{g}$. We also set

$$V^{\mathrm{r}} := V \cap \mathfrak{g}^{\mathrm{r}}$$

for any subset $V \subseteq \mathfrak{g}$.

Let $\langle \cdot, \cdot \rangle : \mathfrak{g} \otimes_{\mathbb{C}} \mathfrak{g} \longrightarrow \mathbb{C}$ denote the Killing form, and write $V^{\perp} \subseteq \mathfrak{g}$ for the annihilator of a subspace $V \subseteq \mathfrak{g}$ under this form. The Killing form is non-degenerate and G-invariant, implying that

$$\mathfrak{g} \longrightarrow \mathfrak{g}^*, \quad x \mapsto \langle x, \cdot \rangle, \quad x \in \mathfrak{g} \tag{2.1}$$

and

$$\mathfrak{g} \oplus \mathfrak{g} \longrightarrow (\mathfrak{g} \oplus \mathfrak{g})^*, \quad (x_1, x_2) \mapsto (\langle x_1, \cdot \rangle, -\langle x_2, \cdot \rangle), \quad (x_1, x_2) \in \mathfrak{g} \oplus \mathfrak{g}$$

define G-module and $(G \times G)$-module isomorphisms, respectively. We use these isomorphisms to freely identify $\mathfrak{g}$ with $\mathfrak{g}^*$ and $\mathfrak{g} \oplus \mathfrak{g}$ with $(\mathfrak{g} \oplus \mathfrak{g})^*$ throughout the paper. The canonical Poisson structure on $\mathfrak{g}^*$ then renders $\mathfrak{g}$ a Poisson variety. This endows the coordinate algebra $\mathbb{C}[\mathfrak{g}] = \mathrm{Sym}(\mathfrak{g}^*)$ with a Poisson bracket, defined as follows:

$$\{f_1, f_2\}(x) = \langle x, [(df_1)_x, (df_2)_x] \rangle$$

for all $f_1, f_2 \in \mathbb{C}[\mathfrak{g}]$ and $x \in \mathfrak{g}$, where $(df_1)_x, (df_2)_x \in \mathfrak{g}^*$ are regarded as elements of $\mathfrak{g}$ via (2.1). We also note that the symplectic leaves of $\mathfrak{g}$ are precisely the adjoint orbits

$$Gx := \{\mathrm{Ad}_g(x) : g \in G\}, \quad x \in \mathfrak{g}.$$

We henceforth fix a principal $\mathfrak{sl}_2$-triple τ, i.e. $\tau = (\xi, h, \eta) \in \mathfrak{g}^{\oplus 3}$ with $\xi, h, \eta \in \mathfrak{g}^{\mathrm{r}}$ and

$$[\xi, \eta] = h, \quad [h, \xi] = 2\xi, \quad \text{and} \quad [h, \eta] = -2\eta.$$

Let $\mathfrak{b} \subseteq \mathfrak{g}$ be the unique Borel subalgebra that contains η. The Cartan subalgebra $\mathfrak{t} := \mathfrak{g}_h$ then satisfies $\mathfrak{t} \subseteq \mathfrak{b}$, and this gives rise to collections of roots $\Phi \subseteq \mathfrak{t}^*$, positive roots $\Phi^+ \subseteq \Phi$, negative roots $\Phi^- = -\Phi^+$, and simple roots $\Pi \subseteq \Phi^+$. Each subset $I \subseteq \Pi$ then determines parabolic subalgebras

$$\mathfrak{p}_I := \mathfrak{b} \oplus \bigoplus_{\alpha \in \Phi_I^-} \mathfrak{g}_\alpha \quad \text{and} \quad \mathfrak{p}_I^- = \mathfrak{b}^- \oplus \bigoplus_{\alpha \in \Phi_I^+} \mathfrak{g}_\alpha,$$

where $\mathfrak{b}^- \subseteq \mathfrak{g}$ is the Borel subalgebra opposite to $\mathfrak{b}$ with respect to $\mathfrak{t}$ and Φ_I^+ (resp. Φ_I^-) is the set of positive (resp. negative) roots in the $\mathbb{Z}$-span of I. Note that

$$\mathfrak{l}_I := \mathfrak{p}_I \cap \mathfrak{p}_I^-$$

is a Levi subalgebra of $\mathfrak{g}$. Let $\mathfrak{u}_I$ and $\mathfrak{u}_I^-$ denote the nilradicals of $\mathfrak{p}_I$ and $\mathfrak{p}_I^-$, respectively, observing that

$$\mathfrak{p}_I = \mathfrak{l}_I \oplus \mathfrak{u}_I \quad \text{and} \quad \mathfrak{p}_I^- = \mathfrak{l}_I \oplus \mathfrak{u}_I^-.$$

We have $\mathfrak{p}_\emptyset = \mathfrak{b}$, $\mathfrak{p}_\emptyset^- = \mathfrak{b}^-$, and $\mathfrak{l}_\emptyset = \mathfrak{t}$, and we adopt the notation

$$\mathfrak{u} := \mathfrak{u}_\emptyset \quad \text{and} \quad \mathfrak{u}^- := \mathfrak{u}_\emptyset^-.$$

We conclude with a brief discussion of invariant theory. To this end, consider the finitely-generated subalgebra of $\mathbb{C}[\mathfrak{g}]$ given by

$$\mathbb{C}[\mathfrak{g}]^G := \{f \in \mathbb{C}[\mathfrak{g}] : f(\mathrm{Ad}_g(x)) = f(x) \text{ for all } g \in G \text{ and } x \in \mathfrak{g}\}.$$

Denote by

$$\chi : \mathfrak{g} \longrightarrow \mathrm{Spec}(\mathbb{C}[\mathfrak{g}]^G)$$

the morphism of affine varieties corresponding to the inclusion $\mathbb{C}[\mathfrak{g}]^G \subseteq \mathbb{C}[\mathfrak{g}]$. This morphism is called the *adjoint quotient* of $\mathfrak{g}$.

Now consider the Slodowy slice

$$\mathcal{S} := \xi + \mathfrak{g}_\eta$$

determined by our principal $\mathfrak{sl}_2$-triple τ. This slice consists of regular elements, and it is a fundamental domain for the adjoint action of G on $\mathfrak{g}^{\mathrm{r}}$ (see [**29**, Theorem 8]). This slice is also called a *Kostant section*, reflecting the fact that

$$\chi\big|_{\mathcal{S}} : \mathcal{S} \longrightarrow \mathrm{Spec}(\mathbb{C}[\mathfrak{g}]^G)$$

is an isomorphism (see [**29**, Theorem 7]). We may therefore consider

$$x_{\mathcal{S}} := (\chi\big|_{\mathcal{S}})^{-1}(\chi(x))$$

for each $x \in \mathfrak{g}$, i.e. $x_{\mathcal{S}}$ is the unique point at which $\mathcal{S}$ meets $\chi^{-1}(\chi(x))$. It is known that

$$\chi^{-1}(\chi(x)) = \overline{Gx_{\mathcal{S}}} \quad \text{and} \quad \chi^{-1}(\chi(x)) \cap \mathfrak{g}^{\mathrm{r}} = Gx_{\mathcal{S}}$$

for all $x \in \mathfrak{g}$, and that

$$\chi^{-1}(\chi(x)) = Gx = Gx_{\mathcal{S}}$$

for all $x \in \mathfrak{g}^{\mathrm{rs}}$ (see [**29**, Theorem 3]).

2.3. Poisson varieties. Let X be a smooth variety with structure sheaf $\mathcal{O}_X$ and tangent bundle TX. Suppose that P is a global section of $\Lambda^2(TX)$, and consider the bracket on $\mathcal{O}_X$ defined by

$$\{f_1, f_2\} := P(df_1 \wedge df_2)$$

for all open subsets $U \subseteq X$ and functions $f_1, f_2 \in \mathcal{O}_X(U)$. We refer to (X, P) as a *Poisson variety* if the aforementioned bracket renders $\mathcal{O}_X$ a sheaf of Poisson algebras. In this case, P is called the *Poisson bivector.* We shall always understand $(X_1 \times X_2, P_1 \oplus (-P_2))$ as being the product of the Poisson varieties (X_1, P_1) and (X_2, P_2). We also note that every symplectic variety is canonically a Poisson variety.

Let (X, P) be a Poisson variety. We may contract P with covectors and realize it as a bundle morphism

$$P : T^*X \longrightarrow TX,$$

whose image is a singular distribution on X. The maximal integral submanifolds of this distribution are called the *symplectic leaves* of X, and each comes equipped with a holomorphic symplectic form. The symplectic form ω_L on a symplectic leaf $L \subseteq X$ is constructed as follows. One defines the *Hamiltonian vector field* of a locally defined function $f \in \mathcal{O}_X$ by

$$H_f := -P(df).$$

The gives rise to the tangent space description

$$T_xL = \{(H_f)_x : f \in \mathcal{O}_X\}$$

for all $x \in L$, and one has

$$(\omega_L)_x((H_{f_1})_x, (H_{f_2})_x) = \{f_1, f_2\}(x)$$

for all $x \in L$ and $f_1, f_2 \in \mathcal{O}_X$ defined near x.

We now briefly discuss the notion of a *log symplectic variety* [**22**–**26**, **34**, **35**], a slight variant of a symplectic variety in the Poisson category. To this end, let (X, P) be a Poisson variety with a unique open dense symplectic leaf. It follows that $\dim X = 2n$ for some non-negative integer n, and one may consider the section P^n of $\Lambda^{2n}(TX)$. One calls (X, P) a *log symplectic variety* if the vanishing locus of P^n is a reduced normal crossing divisor $Y \subseteq X$. In this case, we call Y *the divisor* of (X, P).

2.4. Poisson slices. Let (X, P) be a Poisson variety. A smooth locally closed subvariety $Z \subseteq X$ is called a *Poisson transversal* if

$$T_zX = T_zZ \oplus P_z((T_zZ)^\dagger)$$

for all $z \in Z$, where $(T_zZ)^\dagger \subseteq T_z^*X$ is the annihilator of T_zZ. This decomposition induces an inclusion $T^*Z \subseteq T^*X$, and one has $P(T^*Z) \subseteq TZ \subseteq TX$. The restriction

$$P_Z := P\big|_{T^*Z} : T^*Z \longrightarrow TZ$$

is then a Poisson bivector on Z. We regard all Poisson transversals as coming equipped with the aforementioned Poisson variety structure. It is also worth noting that the preimage of a Poisson transversal under a Poisson variety morphism is itself a Poisson transversal [**21**, Lemma 1].

Now suppose that (X, P, ν) is a Hamiltonian G-variety, i.e. (X, P) is Poisson variety endowed with a Hamiltonian G-action and moment map $\nu : X \longrightarrow \mathfrak{g}$. Let us also recall the Slodowy slice $\mathcal{S} \subseteq \mathfrak{g}$ fixed in Section 2.2. The preimage $\nu^{-1}(\mathcal{S})$ is a Poisson transversal in X and an example of a Poisson slice [**17**]. Some salient properties of this slice appear in the following specialized version of [**17**, Corollary 3.7].

PROPOSITION 2.1. *Suppose that (X, P, ν) is a Hamiltonian G-variety.*

(i) *If (X, P) is symplectic, then $\nu^{-1}(\mathcal{S})$ is a symplectic subvariety of X. The symplectic structure on $\nu^{-1}(\mathcal{S})$ coincides with the Poisson structure that $\nu^{-1}(\mathcal{S})$ inherits as a Poisson transversal.*

(ii) *Suppose that (X, P) is log symplectic with divisor Y, and that Z is an irreducible component of $\nu^{-1}(\mathcal{S})$. The Poisson structure that Z inherits from $\nu^{-1}(\mathcal{S})$ being a Poisson transversal is then log symplectic with divisor $Z \cap Y$.*

REMARK 2.2. Analogous statements hold in the case of a Poisson variety (X, P) endowed with a Hamiltonian $(G \times G)$-action and moment map $\nu : X \longrightarrow \mathfrak{g} \oplus \mathfrak{g}$. The preimage $\nu^{-1}(\mathcal{S} \times \mathcal{S})$ is a Poisson transversal in X, and Proposition 2.1 holds with "$\nu^{-1}(\mathcal{S} \times \mathcal{S})$" in place of "$\nu^{-1}(\mathcal{S})$".

3. Poisson slices and the wonderful compactification

This section considers the implications of Proposition 2.1 for the cases $X = T^*G$ and $X = T^*\overline{G}(\log D)$, where $T^*\overline{G}(\log D)$ is the log cotangent bundle of the wonderful compactification $\overline{G}$.

3.1. The wonderful compactification of G. Recall the Lie-theoretic notation and conventions established in Section 2.2. The adjoint action of $G \times G$ on $\mathfrak{g} \oplus \mathfrak{g}$ is then given by

$$(g_1, g_2) \cdot (x, y) = (\mathrm{Ad}_{g_1}(x), \mathrm{Ad}_{g_2}(y)), \quad (g_1, g_2) \in G \times G, \ (x, y) \in \mathfrak{g} \oplus \mathfrak{g}.$$

This induces an action on the Grassmannian $\mathrm{Gr}(n, \mathfrak{g} \oplus \mathfrak{g})$ of n-dimensional subspaces in $\mathfrak{g} \oplus \mathfrak{g}$, where $n = \dim \mathfrak{g}$. Now consider the point

$$\mathfrak{g}_\Delta := \{(x, x) \colon x \in \mathfrak{g}\} \in \mathrm{Gr}(n, \mathfrak{g} \oplus \mathfrak{g})$$

and set

$$\gamma_g := (g, e) \cdot \mathfrak{g}_\Delta = \{(\mathrm{Ad}_g(x), x) : x \in \mathfrak{g}\} \in \mathrm{Gr}(n, \mathfrak{g} \oplus \mathfrak{g})$$

for each $g \in G$. The map

$$\kappa : G \longrightarrow \mathrm{Gr}(n, \mathfrak{g} \oplus \mathfrak{g}), \quad g \mapsto \gamma_g, \quad g \in G \tag{3.1}$$

is a $(G \times G)$-equivariant locally closed embedding. The closure of its image is a $(G \times G)$-invariant closed subvariety of $\mathrm{Gr}(n, \mathfrak{g} \oplus \mathfrak{g})$, often denoted

$$\overline{G} \subseteq \mathrm{Gr}(n, \mathfrak{g} \oplus \mathfrak{g})$$

and called the *wonderful compactification* of G.

It is known that $\overline{G}$ is smooth [**20**, Proposition 2.14], and that $D := \overline{G} \setminus G$ is a normal crossing divisor [**20**, Theorem 2.22]. One also knows that $\overline{G}$ is stratified into finitely many $(G \times G)$-orbits, indexed by the subsets $I \subseteq \Pi$ [**20**, Theorem 2.22

and Remark 3.9]. To describe the $(G\times G)$-orbit corresponding to $I\subseteq\Pi$, recall the Levi decompositions

$$\mathfrak{p}_I=\mathfrak{l}_I\oplus\mathfrak{u}_I\quad\text{and}\quad\mathfrak{p}_I^-=\mathfrak{l}_I\oplus\mathfrak{u}_I^-$$

from Section 2.2. These allow us to define an n-dimensional subspace of $\mathfrak{g}\oplus\mathfrak{g}$ by

$$\mathfrak{p}_I\times_{\mathfrak{l}_I}\mathfrak{p}_I^-:=\{(x,y)\in\mathfrak{p}_I\oplus\mathfrak{p}_I^-: x\text{ and }y\text{ have the same projection to }\mathfrak{l}_I\}.\tag{3.2}$$

The $(G\times G)$-orbit corresponding to I is then given by

$$(G\times G)(\mathfrak{p}_I\times_{\mathfrak{l}_I}\mathfrak{p}_I^-)\subseteq\overline{G}.$$

The following lemma (cf. [**20**, Remark 3.9]) implies that $\mathfrak{p}_I\times_{\mathfrak{l}_I}\mathfrak{p}_I^-\in\overline{G}$, justifying the inclusion asserted above. This lemma features prominently in later sections.

LEMMA 3.1. *Let $\tilde{T}\subseteq G$ be a maximal torus with Lie algebra $\tilde{\mathfrak{t}}\subseteq\mathfrak{g}$, and consider a one-parameter subgroup $\lambda:\mathbb{C}^\times\longrightarrow\tilde{T}$. Let $\mathfrak{p}$ be the parabolic subalgebra spanned by $\tilde{\mathfrak{t}}$ and the root spaces $\mathfrak{g}_\alpha$ for all roots α of $(\mathfrak{g},\tilde{\mathfrak{t}})$ satisfying $(\alpha,\lambda)\geq 0$, where $(\cdot,\cdot)$ is the pairing between weights and coweights. Let $\mathfrak{l}\subseteq\mathfrak{g}$ be the Levi subalgebra spanned by $\tilde{\mathfrak{t}}$ and all $\mathfrak{g}_\alpha$ with $(\alpha,\lambda)=0$. Write $\mathfrak{p}^-$ for the opposite parabolic, spanned by $\tilde{\mathfrak{t}}$ and those root spaces $\mathfrak{g}_\alpha$ such that $(\alpha,\lambda)\leq 0$. We then have*

$$\lim_{t\longrightarrow\infty}(\lambda(t),e)\cdot\mathfrak{g}_\Delta=\mathfrak{p}\times_{\mathfrak{l}}\mathfrak{p}^-$$

in $\mathrm{Gr}(n,\mathfrak{g}\oplus\mathfrak{g})$, where the right-hand side is defined analogously to (3.2).

PROOF. Choose a Borel subalgebra $\tilde{\mathfrak{b}}\subseteq\mathfrak{g}$ satisfying $\tilde{\mathfrak{t}}\subseteq\tilde{\mathfrak{b}}\subseteq\mathfrak{p}$, and write $\tilde{\Phi}$, $\tilde{\Phi}^+$, and $\tilde{\Pi}$ for the associated sets of roots, positive roots, and simple roots, respectively. The subset $I:=\{\alpha\in\tilde{\Pi}:(\alpha,\lambda)=0\}$ then corresponds to the standard parabolic subalgebra $\mathfrak{p}$. Let $\tilde{\Phi}_I^+\subseteq\tilde{\Phi}^+$ denote the set of positive roots in the $\mathbb{Z}$-span of I. We then have

$$\tilde{\Phi}_I^+=\{\alpha\in\tilde{\Phi}^+:(\alpha,\lambda)=0\}\quad\text{and}\quad\tilde{\Phi}^+\setminus\tilde{\Phi}_I^+=\{\alpha\in\tilde{\Phi}^+:(\alpha,\lambda)>0\}.$$

Choose a non-zero root vector $e_\alpha\in\mathfrak{g}_\alpha$ for each $\alpha\in\tilde{\Phi}$, and fix a basis $\{h_1,\ldots,h_\ell\}$ of $\tilde{\mathfrak{t}}$. It follows that

$$\{(e_\alpha,e_\alpha)\}_{\alpha\in\tilde{\Phi}}\cup\{(h_i,h_i)\}_{i=1}^{\ell}$$

is a basis of $\mathfrak{g}_\Delta$. Now set

$$E_\alpha:=(e_\alpha,e_\alpha),\quad E_\alpha^1:=(e_\alpha,0),\quad\text{and}\quad E_\alpha^2:=(0,e_\alpha)$$

for each $\alpha\in\tilde{\Phi}$, and also write

$$H_i:=(h_i,h_i)$$

for $i=1,\ldots,\ell$. Observe that $(\lambda(t),e)\cdot\mathfrak{g}_\Delta$ then has a basis of

$$\begin{aligned}&\{t^{(\alpha,\lambda)}E_\alpha^1+E_\alpha^2\}_{\alpha\in\tilde{\Phi}^+\setminus\tilde{\Phi}_I^+}\cup\{t^{-(\alpha,\lambda)}E_{-\alpha}^1+E_{-\alpha}^2\}_{\alpha\in\tilde{\Phi}^+\setminus\tilde{\Phi}_I^+}\\&\cup\{E_\beta\}_{\beta\in\tilde{\Phi}_I^+}\cup\{E_{-\beta}\}_{\beta\in\tilde{\Phi}_I^+}\cup\{H_i\}_{i=1}^{\ell}.\end{aligned}$$

The image of $(\lambda(t),e)\cdot\mathfrak{g}_\Delta$ under the Plücker embedding

$$\vartheta:\mathrm{Gr}(n,\mathfrak{g}\oplus\mathfrak{g})\longrightarrow\mathbb{P}(\Lambda^n(\mathfrak{g}\oplus\mathfrak{g})),\quad V\mapsto[\Lambda^nV],\quad V\in\mathrm{Gr}(n,\mathfrak{g}\oplus\mathfrak{g})$$

is therefore

$$\vartheta((\lambda(t),e)\cdot\mathfrak{g}_\Delta)$$
$$=\left[\bigwedge_{\alpha\in\tilde{\Phi}^+\setminus\tilde{\Phi}^+_I}(t^{(\alpha,\lambda)}E^1_\alpha+E^2_\alpha)\wedge\bigwedge_{i=1}^{\ell}H_i\wedge\bigwedge_{\beta\in\tilde{\Phi}^+_I}(E_\beta\wedge E_{-\beta})\wedge\bigwedge_{\alpha\in\tilde{\Phi}^+\setminus\tilde{\Phi}^+_I}(t^{-(\alpha,\lambda)}E^1_{-\alpha}+E^2_{-\alpha})\right].$$

We have

$$\bigwedge_{\alpha\in\tilde{\Phi}^+\setminus\tilde{\Phi}^+_I}(t^{(\alpha,\lambda)}E^1_\alpha+E^2_\alpha)=t^{\sum_{\alpha\in\tilde{\Phi}^+\setminus\tilde{\Phi}^+_I}(\alpha,\lambda)}\Bigg(\bigwedge_{\alpha\in\tilde{\Phi}^+\setminus\tilde{\Phi}^+_I}(E^1_\alpha+t^{-(\alpha,\lambda)}E^2_\alpha)\Bigg),$$

and thus

$$\vartheta((\lambda(t),e)\cdot\mathfrak{g}_\Delta)$$
$$=\left[\bigwedge_{\alpha\in\tilde{\Phi}^+\setminus\tilde{\Phi}^+_I}(E^1_\alpha+t^{-(\alpha,\lambda)}E^2_\alpha)\wedge\bigwedge_{i=1}^{\ell}H_i\wedge\bigwedge_{\beta\in\tilde{\Phi}^+_I}(E_\beta\wedge E_{-\beta})\wedge\bigwedge_{\alpha\in\tilde{\Phi}^+\setminus\tilde{\Phi}^+_I}(t^{-(\alpha,\lambda)}E^1_{-\alpha}+E^2_{-\alpha})\right].$$

All exponents of t appearing in this expression are strictly negative, implying that

$$\lim_{t\longrightarrow\infty}\vartheta((\lambda(t),e)\cdot\mathfrak{g}_\Delta)$$
$$=\left[\bigwedge_{\alpha\in\tilde{\Phi}^+\setminus\tilde{\Phi}^+_I}E^1_\alpha\wedge\bigwedge_{i=1}^{\ell}H_i\wedge\bigwedge_{\beta\in\tilde{\Phi}^+_I}(E_\beta\wedge E_{-\beta})\wedge\bigwedge_{\alpha\in\tilde{\Phi}^+\setminus\tilde{\Phi}^+_I}E^2_{-\alpha}\right]=\vartheta(\mathfrak{p}\times_{\mathfrak{l}}\mathfrak{p}^-).$$

The desired conclusion now follows from the $(G\times G)$-equivariance of the Plücker embedding ϑ. $\square$

3.2. The slices $G\times\mathcal{S}$ and $\overline{G\times\mathcal{S}}$. One can consider the *log cotangent bundle*

$$\pi: T^*\overline{G}(\log D)\longrightarrow\overline{G},$$

i.e. the vector bundle associated with the locally free sheaf of logarithmic one-forms on $(\overline{G},D)$. It turns out that $T^*\overline{G}(\log D)$ is the pullback of the tautological bundle $\mathbb{T}\longrightarrow\mathrm{Gr}(n,\mathfrak{g}\oplus\mathfrak{g})$ to the subvariety $\overline{G}$ (see [**5**, Section 3.1] or [**12**, Example 2.5]). In other words,

$$T^*\overline{G}(\log D)=\{(\gamma,(x,y))\in\overline{G}\times(\mathfrak{g}\oplus\mathfrak{g})\colon(x,y)\in\gamma\}\tag{3.3}$$

and

$$\pi(\gamma,(x,y))=\gamma$$

for all $(\gamma,(x,y))\in T^*\overline{G}(\log D)$.

The variety $T^*\overline{G}(\log D)$ carries a natural log symplectic structure, some aspects of which we now describe. To this end, use the left trivialization and Killing form to identify T^*G with $G\times\mathfrak{g}$. It is straightforward to verify that

$$(g_1,g_2)\cdot(h,x)=(g_1hg_2^{-1},\mathrm{Ad}_{g_2}(x)),\quad(g_1,g_2)\in G\times G,\ (h,x)\in G\times\mathfrak{g}\tag{3.4}$$

defines a Hamiltonian action of $G\times G$ on T^*G, and that

$$\mu=(\mu_L,\mu_R):T^*G\longrightarrow\mathfrak{g}\oplus\mathfrak{g},\quad(g,x)\mapsto(\mathrm{Ad}_g(x),x),\quad(g,x)\in G\times\mathfrak{g}\tag{3.5}$$

is a moment map.

The map

$$\Psi:T^*G\longrightarrow T^*\overline{G}(\log D),\quad(g,x)\mapsto(\gamma_g,(\mathrm{Ad}_g(x),x)),\quad(g,x)\in G\times\mathfrak{g}\tag{3.6}$$

defines an open embedding of T^*G into $T^*\overline{G}(\log D)$, and it fits into a pullback square

$$\begin{array}{ccc} T^*G & \xrightarrow{\Psi} & T^*\overline{G}(\log D) \\ \downarrow & & \downarrow \\ G & \xrightarrow{\kappa} & \overline{G} \end{array}.$$

This embedding is known to define a symplectomorphism from T^*G to the unique open dense symplectic leaf in $T^*\overline{G}(\log D)$ [**5**, Section 3.3]. On the other hand, the following defines a Hamiltonian action of $G \times G$ on $T^*\overline{G}(\log D)$:

$$(g_1, g_2) \cdot (\gamma, (x, y)) := ((g_1, g_2) \cdot \gamma, (\mathrm{Ad}_{g_1}(x), \mathrm{Ad}_{g_2}(y))), \quad (g_1, g_2) \in G \times G,\ (\gamma, (x, y)) \in T^*\overline{G}(\log D), \tag{3.7}$$

where $(g_1, g_2) \cdot \gamma$ refers to the action of $G \times G$ on $\overline{G}$. An associated moment map is given by

$$\overline{\mu} = (\overline{\mu}_L, \overline{\mu}_R) : T^*\overline{G}(\log D) \longrightarrow \mathfrak{g} \oplus \mathfrak{g}, \quad (\gamma, (x, y)) \mapsto (x, y), \quad (\gamma, (x, y)) \in T^*\overline{G}(\log D)$$

(see [**5**, Section 3.2] and [**12**, Example 2.5]). The open embedding Ψ is then $(G\times G)$-equivariant and satisfies

$$\Psi^*\overline{\mu} = \mu.$$

We now discuss the Poisson slices

$$G \times \mathcal{S} = \mu_R^{-1}(\mathcal{S}) \subseteq T^*G \quad \text{and} \quad \overline{G \times \mathcal{S}} := \overline{\mu}_R^{-1}(\mathcal{S}) \subseteq T^*\overline{G}(\log D),$$

using [**17**, Sections 3.2 and 3.2] as our reference. The former slice is symplectic, while the latter is log symplectic. One also knows that $\Psi : T^*G \longrightarrow T^*\overline{G}(\log D)$ restricts to a G-equivariant open embedding

$$\Psi\big|_{G\times\mathcal{S}} : G \times \mathcal{S} \longrightarrow \overline{G \times \mathcal{S}}, \tag{3.8}$$

and that this embedding is a symplectomorphism onto the unique open dense symplectic leaf in $\overline{G \times \mathcal{S}}$. We also note that G acts in a Hamiltonian fashion on $G \times \mathcal{S}$ (resp. $\overline{G \times \mathcal{S}}$) via (3.4) (resp. (3.7)) as the subgroup $G = G \times \{e\} \subseteq G \times G$. The maps $\mu_L : T^*G \longrightarrow \mathfrak{g}$ and $\overline{\mu}_L : T^*\overline{G}(\log D) \longrightarrow \mathfrak{g}$ then restrict to moment maps

$$\mu_{\mathcal{S}} := \mu_L\Big|_{G\times\mathcal{S}} : G \times \mathcal{S} \longrightarrow \mathfrak{g} \quad \text{and} \quad \overline{\mu}_{\mathcal{S}} := \overline{\mu}_L\Big|_{\overline{G\times\mathcal{S}}} : \overline{G \times \mathcal{S}} \longrightarrow \mathfrak{g},$$

and we have a commutative diagram

$$\begin{array}{ccc} G \times \mathcal{S} & \xrightarrow{\Psi|_{G\times\mathcal{S}}} & \overline{G \times \mathcal{S}} \\ & {\scriptstyle \mu_{\mathcal{S}}} \searrow \quad \swarrow {\scriptstyle \overline{\mu}_{\mathcal{S}}} & \\ & \mathfrak{g} & \end{array}. \tag{3.9}$$

An explicit description of $\overline{G \times \mathcal{S}}$ is obtained as follows. One begins by noting that

$$\overline{G \times \mathcal{S}} = \overline{\mu}_R^{-1}(\mathcal{S}) = \{(\gamma, (x, y)) \in \overline{G} \times (\mathfrak{g} \oplus \mathfrak{g}) \colon (x, y) \in \gamma \text{ and } y \in \mathcal{S}\}.$$

On the other hand, recall the adjoint quotient

$$\chi : \mathfrak{g} \longrightarrow \mathrm{Spec}(\mathbb{C}[\mathfrak{g}]^G)$$

and the associated concepts discussed in Section 2.2. The image of $\overline{\mu}$ is known to be

$$\{(x,y)\in\mathfrak{g}\oplus\mathfrak{g}\colon \chi(x)=\chi(y)\}$$

(see [**5**, Proposition 3.4]), and this implies the simplified description

$$\overline{G\times\mathcal{S}}=\{(\gamma,(x,x_{\mathcal{S}})):\gamma\in\overline{G},\ x\in\mathfrak{g},\ \text{and}\ (x,x_{\mathcal{S}})\in\gamma\}. \tag{3.10}$$

REMARK 3.2. Recall the bundle projection map $\pi : T^*\overline{G}(\log D)\longrightarrow\overline{G}$, and set

$$\pi_{\mathcal{S}}:=\pi\big|_{\overline{G\times\mathcal{S}}}:\overline{G\times\mathcal{S}}\longrightarrow\overline{G}.$$

The description (3.10) implies that the product map

$$(\pi_{\mathcal{S}},\overline{\mu}_{\mathcal{S}}):\overline{G\times\mathcal{S}}\longrightarrow\overline{G}\times\mathfrak{g},\quad(\gamma,(x,x_{\mathcal{S}}))\mapsto(\gamma,x),\quad(\gamma,(x,x_{\mathcal{S}}))\in\overline{G\times\mathcal{S}}$$

is a closed embedding. We thereby obtain a commutative diagram

$$\begin{array}{ccc}\overline{G\times\mathcal{S}} & \longrightarrow & \overline{G}\times\mathfrak{g}\\ & {\scriptstyle\overline{\mu}_{\mathcal{S}}}\searrow \quad \swarrow & \\ & \mathfrak{g} & \end{array},$$

where $\overline{G}\times\mathfrak{g}\longrightarrow\mathfrak{g}$ is projection to the second factor. We conclude that $\overline{\mu}_{\mathcal{S}}$ has projective fibres, so that (3.9) realizes $\overline{\mu}_{\mathcal{S}}$ as a fibrewise compactification of $\mu_{\mathcal{S}}$. It also follows that

$$\overline{\mu}_{\mathcal{S}}^{-1}(x)\longrightarrow\{\gamma\in\overline{G}:(x,x_{\mathcal{S}})\in\gamma\},\quad(\gamma,(x,x_{\mathcal{S}}))\mapsto\gamma,\quad(\gamma,(x,x_{\mathcal{S}}))\in\overline{\mu}_{\mathcal{S}}^{-1}(x)$$

is a variety isomorphism for each $x\in\mathfrak{g}$.

4. Relation to the standard family of Hessenberg varieties

4.1. The standard family of Hessenberg varieties. Recall the notation and conventions established in Section 2.2, and let $B\subseteq G$ be the Borel subgroup with Lie algebra $\mathfrak{b}$. Suppose that $H\subseteq\mathfrak{g}$ is a *Hessenberg subspace*, i.e. a B-invariant subspace of $\mathfrak{g}$ that contains $\mathfrak{b}$. Let $G\times B$ act on $G\times H$ via

$$(g,b)\cdot(h,x):=(ghb^{-1},\mathrm{Ad}_b(x)),\quad(g,b)\in G\times B,\ (h,x)\in G\times H,$$

and consider the resulting smooth G-variety

$$G\times_B H:=(G\times H)/B.$$

Write $[g:x]$ for the equivalence class of $(g,x)\in G\times H$ in $G\times_B H$. The variety $G\times_B H$ is naturally Poisson, and its G-action is Hamiltonian with moment map

$$\nu_H:G\times_B H\longrightarrow\mathfrak{g},\quad[g:x]\mapsto\mathrm{Ad}_g(x),\quad[g:x]\in G\times_B H$$

(see [**5**, Section 4]). Given any $x\in\mathfrak{g}$, one writes

$$\mathrm{Hess}(x,H):=\nu_H^{-1}(x)$$

and calls this fibre the *Hessenberg variety* associated to H and x. The Poisson moment map ν_H is thereby called the *family of Hessenberg varieties* associated to H.

REMARK 4.1. Note that $G \times_B H$ is the total space of a G-equivariant vector bundle over G/B, and that the bundle projection map is

$$\pi_H : G \times_B H \longrightarrow G/B, \quad [g : x] \mapsto [g], \quad [g : x] \in G \times_B H. \tag{4.1}$$

The map

$$(\pi_H, \nu_H) : G \times_B H \longrightarrow G/B \times \mathfrak{g}$$

is then a closed embedding. We also have a commutative diagram

$$\begin{array}{ccc} G \times_B H & \xrightarrow{(\pi_H, \nu_H)} & G/B \times \mathfrak{g} \\ & {\scriptstyle \nu_H} \searrow \quad \swarrow & \\ & \mathfrak{g} & \end{array}, \tag{4.2}$$

where $G/B \times \mathfrak{g} \longrightarrow \mathfrak{g}$ is projection to the second factor. It follows that the fibres of ν_H are projective, and that π_H restricts to a closed embedding

$$\mathrm{Hess}(x, H) \hookrightarrow G/B$$

for each $x \in \mathfrak{g}$. One may thereby regard Hessenberg varieties as closed subvarieties of G/B.

In what follows, we restrict our attention to the so-called *standard family* of Hessenberg varieties. This is defined to be the family

$$\nu := \nu_{\mathfrak{m}} : G \times_B \mathfrak{m} \longrightarrow \mathfrak{g}$$

associated to the *standard Hessenberg subspace*

$$\mathfrak{m} := [\mathfrak{u}, \mathfrak{u}]^{\perp} = \mathfrak{b} \oplus \bigoplus_{\alpha \in \Pi} \mathfrak{g}_{-\alpha}. \tag{4.3}$$

To study this family in more detail, we note the following consequences of the setup in Section 2.2:

$$\xi = \sum_{\alpha \in \Pi} e_{-\alpha} \quad \text{and} \quad \mathfrak{g}_{\eta} \subseteq \mathfrak{u},$$

where

$$e_{-\alpha} \in \mathfrak{g}_{-\alpha} \setminus \{0\}$$

for all $\alpha \in \Pi$. These considerations imply that $\mathcal{S} \subseteq \mathfrak{m}$, allowing one to define the map

$$\rho : G \times \mathcal{S} \longrightarrow G \times_B \mathfrak{m}, \quad (g, x) \mapsto [g : x], \quad (g, x) \in G \times \mathcal{S}.$$

Let us also consider the G-invariant open dense subvariety

$$G \times_B \mathfrak{m}^{\times} \subseteq G \times_B \mathfrak{m},$$

where $\mathfrak{m}^{\times} \subseteq \mathfrak{m}$ is the B-invariant open subvariety defined by

$$\mathfrak{m}^{\times} := \mathfrak{b} + \sum_{\alpha \in \Pi} (\mathfrak{g}_{-\alpha} \setminus \{0\}) := \left\{ x + \sum_{\alpha \in \Pi} c_{\alpha} e_{-\alpha} : x \in \mathfrak{b} \text{ and } c_{\alpha} \in \mathbb{C}^{\times} \text{ for all } \alpha \in \Pi \right\}.$$

One then has the following consequence of [**2**, Theorem 41] and [**5**, Section 4.2 and Theorem 4.16].

PROPOSITION 4.2. *The Poisson variety $G \times_B \mathfrak{m}$ is log symplectic with $G \times_B \mathfrak{m}^{\times}$ as its unique open dense symplectic leaf. The map ρ is a G-invariant symplectomorphism onto the leaf $G \times_B \mathfrak{m}^{\times}$.*

4.2. Some toric geometry. This section uses the techniques of toric geometry to compare the fibres $\nu^{-1}(x) = \mathrm{Hess}(x, \mathfrak{m})$ and $\overline{\mu}_{\mathcal{S}}^{-1}(x)$ over regular semisimple elements $x \in \mathfrak{g}^{\mathrm{rs}}$. Many of the underlying ideas appear in [**5**] and [**20**]. We therefore do not regard this section as containing any original material.

We begin by observing that G_x acts on the fibres of ν and $\overline{\mu}_{\mathcal{S}}$ over x for all $x \in \mathfrak{g}$. This leads to the following two lemmas, parts of which are well-known. To this end, recall the notation and discussion from Section 2.2.

LEMMA 4.3. *Suppose that $x \in \mathfrak{g}^r$. The following statements hold.*

(i) *There is a unique open dense orbit of G_x in $\mathrm{Hess}(x, \mathfrak{m})$.*
(ii) *The group G_x acts freely on the above-mentioned orbit.*
(iii) *If $g \in G$ satisfies $x = \mathrm{Ad}_g(x_{\mathcal{S}})$, then $[g : x_{\mathcal{S}}]$ belongs to the above-mentioned G_x-orbit.*
(iv) *If $x \in \mathfrak{g}^{rs}$, then $\mathrm{Hess}(x, \mathfrak{m})$ is a smooth, projective, toric G_x-variety.*

PROOF. Let $g \in G$ be such that $x = \mathrm{Ad}_g(x_{\mathcal{S}})$. Our first observation is that

$$\nu([g : x_{\mathcal{S}}]) = \mathrm{Ad}_g(x_{\mathcal{S}}) = x,$$

i.e. $[g : x_{\mathcal{S}}] \in \mathrm{Hess}(x, \mathfrak{m})$. At the same time, Corollaries 3 and 14 in [**33**] imply that $\mathrm{Hess}(x, \mathfrak{m})$ is irreducible and ℓ-dimensional. Claims (i), (ii), and (iii) would therefore follow from our showing the G_x-stabilizer of $[g : x_{\mathcal{S}}]$ to be trivial.

Suppose that $h \in G_x$ is such that $[hg : x_{\mathcal{S}}] = [g : x_{\mathcal{S}}]$. It follows that $(hgb^{-1}, \mathrm{Ad}_b(x_{\mathcal{S}})) = (g, x_{\mathcal{S}})$ for some $b \in B$. Since the B-stabilizer of every point in $\mathcal{S}$ is trivial (see [**5**, Lemma 4.9]), we must have $b = e$. This yields the identity $hg = g$, or equivalently $h = e$. In light of the conclusion reached in the previous paragraph, Claims (i), (ii), and (iii) hold.

Claim (iv) is well-known and follows from Theorems 6 and 11 in [**19**]. □

Let us also recall the notation γ_g defined in Section 3.1.

LEMMA 4.4. *Suppose that $x \in \mathfrak{g}^r$. The following statements hold.*

(i) *There is a unique open dense orbit of G_x in $\overline{\mu}_{\mathcal{S}}^{-1}(x)$.*
(ii) *The group G_x acts freely on the above-mentioned orbit.*
(iii) *If $g \in G$ satisfies $x = \mathrm{Ad}_g(x_{\mathcal{S}})$, then $(\gamma_g, (x, x_{\mathcal{S}}))$ belongs to the above-mentioned G_x-orbit.*
(iv) *If $x \in \mathfrak{g}^{rs}$, then $\overline{\mu}_{\mathcal{S}}^{-1}(x)$ is a smooth, projective, toric G_x-variety.*

PROOF. The moment map $\overline{\mu}_{\mathcal{S}}$ is G-equivariant, so that acting by the element g^{-1} defines a variety isomorphism

$$\overline{\mu}_{\mathcal{S}}^{-1}(x) \xrightarrow{\cong} \overline{\mu}_{\mathcal{S}}^{-1}(x_{\mathcal{S}}). \tag{4.4}$$

Since the latter variety is given by

$$\overline{\mu}_{\mathcal{S}}^{-1}(x_{\mathcal{S}}) = \{(\gamma, (x_{\mathcal{S}}, x_{\mathcal{S}})) : \gamma \in \overline{G} \text{ and } (x_{\mathcal{S}}, x_{\mathcal{S}}) \in \gamma\}, \tag{4.5}$$

we can use [**5**, Corollary 3.12] and conclude that $\overline{\mu}_{\mathcal{S}}^{-1}(x_{\mathcal{S}})$ is irreducible and ℓ-dimensional. It follows that $\overline{\mu}_{\mathcal{S}}^{-1}(x)$ is irreducible and ℓ-dimensional. As with the proof of our previous lemma, Claims (i), (ii), and (ii) would now follow from knowing $(\gamma_g, (x, x_{\mathcal{S}}))$ to have a trivial G_x-stabilizer. This is established via a straightforward calculation.

To verify Claim (iv), recall the isomorphism (4.4). This isomorphism implies that $\overline{\mu}_{\mathcal{S}}^{-1}(x)$ is a smooth, projective, toric G_x-variety if and only if $\overline{\mu}_{\mathcal{S}}^{-1}(x_{\mathcal{S}})$ is a smooth, projective, toric $g^{-1}G_xg = G_{x_S}$-variety. The latter condition holds because of (4.5), [**5**, Corollary 3.12], and the fact that the closure of G_{x_S} in $\overline{G}$ is a smooth, projective, toric G_{x_S}-variety [**20**, Remark 4.5]. Our proof is therefore complete. $\square$

Our next two lemmas study the G_x-fixed point sets

$$\mathrm{Hess}(x,\mathfrak{m})^{G_x} \subseteq \mathrm{Hess}(x,\mathfrak{m}) \quad \text{and} \quad \overline{\mu}_S^{-1}(x)^{G_x} \subseteq \overline{\mu}_S^{-1}(x)$$

for each $x \in \mathfrak{g}^{\mathrm{rs}}$. To formulate these results, let $\mathcal{B}$ denote the flag variety of all Borel subalgebras in $\mathfrak{g}$ and set

$$\mathcal{B}_x := \{\widetilde{\mathfrak{b}} \in \mathcal{B} : x \in \widetilde{\mathfrak{b}}\}$$

for each $x \in \mathfrak{g}$. Recall that

$$G/B \longrightarrow \mathcal{B}, \quad [g] \mapsto \mathrm{Ad}_g(\mathfrak{b}), \quad [g] \in G/B \tag{4.6}$$

defines a G-equivariant variety isomorphism. Let us also consider the map

$$\pi_{\mathfrak{m}} : G \times_B \mathfrak{m} \longrightarrow G/B, \quad [g:x] \mapsto [g], \quad [g:x] \in G \times_B \mathfrak{m}$$

and its composition with (4.6), i.e.

$$\varpi_{\mathfrak{m}} : G \times_B \mathfrak{m} \longrightarrow \mathcal{B}, \quad [g:x] \mapsto \mathrm{Ad}_g(\mathfrak{b}), \quad [g:x] \in G \times_B \mathfrak{m}.$$

LEMMA 4.5. *If $x \in \mathfrak{g}^{rs}$, then there is a canonical bijection*

$$\mathcal{B}_x \longrightarrow \mathrm{Hess}(x,\mathfrak{m})^{G_x}, \quad \widetilde{\mathfrak{b}} \mapsto z(\widetilde{\mathfrak{b}}), \quad \widetilde{\mathfrak{b}} \in \mathcal{B}_x$$

satisfying

$$\widetilde{\mathfrak{b}} = \varpi_{\mathfrak{m}}(z(\widetilde{\mathfrak{b}}))$$

for all $\widetilde{\mathfrak{b}} \in \mathcal{B}_x$.

PROOF. Recall that $\pi_{\mathfrak{m}}$ restricts to a closed embedding

$$\mathrm{Hess}(x,\mathfrak{m}) \longrightarrow G/B$$

(see Remark 4.1). Composing this embedding with (4.6), we deduce that $\varpi_{\mathfrak{m}}$ restricts to a closed embedding

$$\mathrm{Hess}(x,\mathfrak{m}) \longrightarrow \mathcal{B}. \tag{4.7}$$

This embedding is G_x-equivariant, implying that it restricts to an injection

$$\mathrm{Hess}(x,\mathfrak{m})^{G_x} \longrightarrow \mathcal{B}^{G_x} = \mathcal{B}_x. \tag{4.8}$$

We claim that (4.8) is also surjective. To this end, suppose that $\widetilde{\mathfrak{b}} \in \mathcal{B}_x$. We then have $\widetilde{\mathfrak{b}} = \mathrm{Ad}_g(\mathfrak{b})$ for some $g \in G$. It follows that $x = \mathrm{Ad}_g(y)$ for some $y \in \mathfrak{b}$, so that we have a point $[g:y] \in \mathrm{Hess}(x,\mathfrak{m})$. The image of this point under (4.7) is $\widetilde{\mathfrak{b}}$. Noting that $\widetilde{\mathfrak{b}} \in \mathcal{B}_x = \mathcal{B}^{G_x}$ and that (4.7) is injective and G_x-equivariant, we conclude that $[g:y] \in \mathrm{Hess}(x,\mathfrak{m})^{G_x}$. This establishes that (4.8) is surjective, i.e. that it is a bijection. The bijection advertised in the statement of the lemma is obtained by inverting (4.8). $\square$

If $x \in \mathfrak{g}^{\mathrm{rs}}$, then the Cartan subalgebra $\mathfrak{g}_x$ satisfies $\mathfrak{g}_x \subseteq \widetilde{\mathfrak{b}}$ for all $\widetilde{\mathfrak{b}} \in \mathcal{B}_x$ (see [**14**, Lemma 3.1.4]). This fact gives rise to an opposite Borel subalgebra $\widetilde{\mathfrak{b}}^- \in \mathcal{B}_x$ for all $\widetilde{\mathfrak{b}} \in \mathcal{B}_x$. We may therefore define

$$\theta(\widetilde{\mathfrak{b}}) := \widetilde{\mathfrak{b}} \times_{\mathfrak{g}_x} \widetilde{\mathfrak{b}}^-$$

analogously to (3.2).

LEMMA 4.6. *Suppose that $x \in \mathfrak{g}^{rs}$ and let $g \in G$ be such that $x = \mathrm{Ad}_g(x_{\mathcal{S}})$. We then have a bijection defined by*

$$\mathcal{B}_x \longrightarrow \overline{\mu}_{\mathcal{S}}^{-1}(x)^{G_x}, \quad \widetilde{\mathfrak{b}} \mapsto \Big((e, g^{-1}) \cdot \theta(\widetilde{\mathfrak{b}}), (x, x_{\mathcal{S}})\Big), \quad \widetilde{\mathfrak{b}} \in \mathcal{B}_x.$$

PROOF. Note that conjugation by g^{-1} defines a bijection $\phi : \mathcal{B}_x \longrightarrow \mathcal{B}_{x_{\mathcal{S}}}$. At the same time, consider the automorphism of $\overline{G \times \mathcal{S}}$ through which the element g^{-1} acts. This automorphism restricts to a bijection

$$\overline{\mu}_{\mathcal{S}}^{-1}(x)^{G_x} \xrightarrow{\cong} \overline{\mu}_{\mathcal{S}}^{-1}(x_{\mathcal{S}})^{G_{x_{\mathcal{S}}}}.$$

It also sends

$$\Big((e, g^{-1}) \cdot \theta(\widetilde{\mathfrak{b}}), (x, x_{\mathcal{S}})\Big) \mapsto (\theta(\phi(\widetilde{\mathfrak{b}})), (x_{\mathcal{S}}, x_{\mathcal{S}}))$$

for all $\widetilde{\mathfrak{b}} \in \mathcal{B}_x$, where the subspaces $\{\theta(\widetilde{\mathfrak{b}})\}_{\widetilde{\mathfrak{b}} \in \mathcal{B}_{x_{\mathcal{S}}}}$ are defined analogously to the $\{\theta(\widetilde{\mathfrak{b}})\}_{\widetilde{\mathfrak{b}} \in \mathcal{B}_x}$. It therefore suffices to prove that

$$\mathcal{B}_{x_{\mathcal{S}}} \longrightarrow \overline{\mu}_{\mathcal{S}}^{-1}(x_{\mathcal{S}})^{G_{x_{\mathcal{S}}}}, \quad \widetilde{\mathfrak{b}} \mapsto (\theta(\widetilde{\mathfrak{b}}), (x_{\mathcal{S}}, x_{\mathcal{S}})), \quad \widetilde{\mathfrak{b}} \in \mathcal{B}_{x_{\mathcal{S}}}$$

defines a bijection. In other words, it suffices to prove our lemma under the assumption that $x \in \mathcal{S} \cap \mathfrak{g}^{\mathrm{rs}}$ and $g = e$.

Let us make the assumption indicated above. The fibre $\overline{\mu}_{\mathcal{S}}^{-1}(x)$ is then given by

$$\overline{\mu}_{\mathcal{S}}^{-1}(x) = \{(\gamma, (x, x)) : \gamma \in \overline{G} \text{ and } (x, x) \in \gamma\}.$$

It now follows from [**5**, Corollary 3.12] that

$$\overline{G_x} \longrightarrow \overline{\mu}_{\mathcal{S}}^{-1}(x), \quad \gamma \mapsto (\gamma, (x, x)), \quad \gamma \in \overline{G_x}$$

defines an isomorphism of varieties, where $\overline{G_x}$ denotes the closure of G_x in $\overline{G}$. This isomorphism is G_x-equivariant if one lets G_x act on $\overline{G_x}$ by restricting the $(G \times G)$-action on $\overline{G}$ to $G_x = G_x \times \{e\} \subseteq G \times G$. We therefore have

$$\overline{\mu}_{\mathcal{S}}^{-1}(x)^{G_x} = \{(\gamma, (x, x)) : \gamma \in (\overline{G_x})^{G_x}\}. \tag{4.9}$$

At the same time, we will prove that

$$(\overline{G_x})^{G_x} = \{\theta(\widetilde{\mathfrak{b}}) : \widetilde{\mathfrak{b}} \in \mathcal{B}_x\} \tag{4.10}$$

in Lemma 4.7. Our current lemma now follows from (4.9), (4.10), and the conclusion of the previous paragraph. □

LEMMA 4.7. *We have*

$$(\overline{G_x})^{G_x} = \{\theta(\widetilde{\mathfrak{b}}) : \widetilde{\mathfrak{b}} \in \mathcal{B}_x\}$$

for all $x \in \mathfrak{g}^{rs}$, where G_x acts on $\overline{G_x}$ as the subgroup $G_x = G_x \times \{e\} \subseteq G \times G$ via the $(G \times G)$-action on $\overline{G}$.

PROOF. It suffices to assume that $x \in \mathfrak{t}^{\mathrm{r}}$, so that $G_x = T$. Now suppose that $\widetilde{\mathfrak{b}} \in \mathcal{B}_x$. Lemma 3.1 implies that

$$\theta(\widetilde{\mathfrak{b}}) = \lim_{t \longrightarrow \infty} (\lambda(t), e) \cdot \mathfrak{g}_{\Delta}$$

for a suitable one-parameter subgroup $\lambda : \mathbb{C}^{\times} \longrightarrow T$, while we observe that

$$(\lambda(t), e) \cdot \mathfrak{g}_{\Delta} \in \kappa(T) \subseteq \overline{G}$$

for all $t \in \mathbb{C}^\times$. It follows that $\theta(\widetilde{\mathfrak{b}}) \in \overline{T}$. A direct calculation establishes that $\theta(\widetilde{\mathfrak{b}})$ is a T-fixed point, yielding the inclusion

$$\{\theta(\widetilde{\mathfrak{b}}) : \widetilde{\mathfrak{b}} \in \mathcal{B}_x\} \subseteq \overline{T}^T.$$

To establish the opposite inclusion, suppose that $\gamma \in (\overline{T})^T$. Using the $(G\times G)$-orbit decomposition of $\overline{G}$ from Section 3.1, we may find $I \subseteq \Pi$ and $g_1, g_2 \in G$ such that

$$\gamma = (g_1, g_2) \cdot \mathfrak{p}_I \times_{\mathfrak{l}_I} \mathfrak{p}_I^-.$$

We will first show that g_1 can be taken to lie in $N_G(T)$ without the loss of generality. Note that $\widetilde{\mathfrak{b}} := \mathrm{Ad}_{g_1}(\mathfrak{b})$ will then be a Borel subalgebra containing x. We will then explain that $\gamma = \theta(\widetilde{\mathfrak{b}})$, completing the proof. In what follows, we will need the following description of the $(G \times G)$-stabilizer of $\mathfrak{p}_I \times_{\mathfrak{l}_I} \mathfrak{p}_I^- \in \overline{G}$:

$$(G \times G)_I := \{(l_1u, l_2v) \in L_IU_I \times L_IU_I^- : l_1l_2^{-1} \in Z(L_I)\}, \tag{4.11}$$

where $L_I \subseteq G$ is the Levi subgroup with Lie algebra $\mathfrak{l}_I$, $Z(L_I)$ is the centre of L_I, and $U_I \subseteq G$ (resp. $U_I^- \subseteq G$) is the unipotent subgroup with Lie algebra $\mathfrak{u}_I$ (resp. $\mathfrak{u}_I^-$) [**20**, Proposition 2.25].

Since γ is fixed by T, we have

$$(t, e) \cdot \gamma = \gamma$$

for all $t \in T$. It follows that

$$(tg_1, g_2) \cdot \mathfrak{p}_I \times_{\mathfrak{l}_I} \mathfrak{p}_I^- = (g_1, g_2) \cdot \mathfrak{p}_I \times_{\mathfrak{l}_I} \mathfrak{p}_I^-,$$

i.e.

$$(g_1^{-1}tg_1, e) \cdot \mathfrak{p}_I \times_{\mathfrak{l}_I} \mathfrak{p}_I^- = \mathfrak{p}_I \times_{\mathfrak{l}_I} \mathfrak{p}_I^-.$$

We deduce that $(g_1^{-1}tg_1, e) \in (G \times G)_I$, which by (4.11) implies that

$$g_1^{-1}tg_1 \in U_IZ(L_I)$$

for all $t \in T$. In other words,

$$g_1^{-1}Tg_1 \subseteq U_IZ(L_I) \subseteq P_I,$$

where $P_I \subseteq G$ is the parabolic subgroup with Lie algebra $\mathfrak{p}_I$. Noting that $g_1^{-1}Tg_1$ and T are maximal tori in P_I, we can find a $p \in P_I$ such that

$$g_1^{-1}Tg_1 = pTp^{-1}.$$

This shows that $(g_1p)^{-1}Tg_1p = T$, so that $g_1p \in N_G(T)$. Now use the decomposition $P_I = L_IU_I$ to write $p = lu$ with $l \in L_I$ and $u \in U_I$. Equation (4.11) then tells us that $(p, l) \in (G \times G)_I$, yielding

$$\gamma = (g_1, g_2) \cdot \mathfrak{p}_I \times_{\mathfrak{l}_I} \mathfrak{p}_I^- = (g_1p, g_2l) \cdot \mathfrak{p}_I \times_{\mathfrak{l}_I} \mathfrak{p}_I^-.$$

We may therefore take $g_1 \in N_G(T)$ without the loss of generality.

Now we show that $I = \emptyset$. An argument given above establishes that

$$T = g_1^{-1}Tg_1 \subseteq U_IZ(L_I),$$

i.e. $T \subseteq U_IZ(L_I)$. Let $t \in T$ be arbitrary and write $t = vz$ with $v \in U_I$ and $z \in Z(L_I)$. Since T contains $Z(L_I)$, we have $z \in T$ and $v = tz^{-1} \in T \cap U_I = \{e\}$. We conclude that $v = e$ and $t \in Z(L_I)$. This shows that $T \subseteq Z(L_I)$, which can only happen if $T = Z(L_I)$. One deduces that $I = \emptyset$, and this yields $\mathfrak{p}_I = \mathfrak{b}$, $\mathfrak{p}_I^- = \mathfrak{b}^-$, $\mathfrak{l}_I = \mathfrak{t}$, and

$$\gamma = (g_1, g_2) \cdot \mathfrak{b} \times_{\mathfrak{t}} \mathfrak{b}^-.$$

We now prove that $g_2 = g_1 \in N_G(T)$ without the loss of generality. Our first observation is that

$$(\mathrm{Ad}_{g_1^{-1}}(x), \mathrm{Ad}_{g_2^{-1}}(x)) \in \mathfrak{b} \times_{\mathfrak{t}} \mathfrak{b}^-,$$

as $(x,x) \in \gamma$. Since $g_1 \in N_G(T)$, we have $\mathrm{Ad}_{g_1^{-1}}(x) \in \mathfrak{t}$. These last two sentences then force

$$\mathrm{Ad}_{g_1^{-1}}(x) - \mathrm{Ad}_{g_2^{-1}}(x) \in \mathfrak{u}^-$$

to hold. We also note that $\mathrm{Ad}_{g_1}^{-1}(x) \in \mathfrak{t}$ is regular, which by [**14**, Lemma 3.1.44] implies that $\mathrm{Ad}_{g_1}^{-1}(x) + \mathfrak{u}^-$ is a U^--orbit. We can therefore find $u_- \in U^-$ such that

$$\mathrm{Ad}_{g_2^{-1}}(x) = \mathrm{Ad}_{u_- g_1^{-1}}(x),$$

i.e. $(g_2 u_-) g_1^{-1} \in G_x = T$. This shows that $g_2 u_- \in T g_1 = g_1 T$, where we have used the fact that $g_1 \in N_G(T)$. We may therefore write $g_2 u_- = g_1 t$ for some $t \in T$, i.e.

$$g_1 = g_2 u_- t^{-1}.$$

Equation (4.11) also implies that $(e, u_- t^{-1}) \in (G \times G)_{\emptyset}$, giving $(e, u_- t^{-1}) \cdot \mathfrak{b} \times_{\mathfrak{t}} \mathfrak{b}^- = \mathfrak{b} \times_{\mathfrak{t}} \mathfrak{b}^-$. This in turn implies that

$$\gamma = (g_1, g_2) \cdot \mathfrak{b} \times_{\mathfrak{t}} \mathfrak{b}^- = (g_1, g_2 u_- t^{-1}) \cdot \mathfrak{b} \times_{\mathfrak{t}} \mathfrak{b}^- = (g_1, g_1) \cdot \mathfrak{b} \times_{\mathfrak{t}} \mathfrak{b}^- = \theta(\widetilde{\mathfrak{b}}),$$

where $\widetilde{\mathfrak{b}} = \mathrm{Ad}_{g_1}(\mathfrak{b})$. Our proof is therefore complete. □

The preceding results have implications for the toric geometries of $\mathrm{Hess}(x, \mathfrak{m})$ and $\overline{\mu}_S^{-1}(x)$, where $x \in \mathfrak{g}^{\mathrm{rs}}$. The following elementary lemma is needed to realize these implications.

LEMMA 4.8. *Suppose that S is a complex torus, and that X is a smooth, projective, toric S-variety with finitely many S-fixed points. Let $\lambda : \mathbb{C}^\times \longrightarrow S$ be a coweight of S, and assume that $(\alpha, \lambda) \neq 0$ for all $y \in X^S$ and all weights α of the isotropy S-representation $T_y X$. Assume that x belongs to the unique open dense orbit S-orbit in X. We have*

$$\lim_{t \longrightarrow \infty} \lambda(t) \cdot x = z$$

for some $z \in X^S$ satisfying $(\alpha, \lambda) < 0$ for all S-weights α of $T_z X$.

PROOF. Let $\mathbb{C}^\times$ act on X through λ. If $y \in X^S$, then $T_y Y$ is an isotropy representation of both S and $\mathbb{C}^\times$. Each weight of the $\mathbb{C}^\times$-representation is given by (α, λ) for a suitable weight α of the S-representation. Since $(\alpha, \lambda) \neq 0$ for all $y \in X^S$ and S-weights α of $T_y X$, the previous sentence implies that $X^S = X^{\mathbb{C}^\times}$.

Now note that $\lim_{t \longrightarrow \infty} \lambda(t) \cdot x$ exists and coincides with a point $z \in X^{\mathbb{C}^\times} = X^S$ (e.g. by [**14**, Lemma 2.4.1]). The previous paragraph explains that each $\mathbb{C}^\times$-weight of $T_z X$ takes the form (α, λ), where α is an S-weight of $T_z X$. General facts about Białynicki-Birula decompositions (e.g. [**14**, Theorem 2.4.3]) now imply that $(\alpha, \lambda) < 0$ for all S-weights α of $T_z X$. □

Suppose that $x \in \mathfrak{g}^{\mathrm{rs}}$. Recall that a coweight $\lambda : \mathbb{C}^\times \longrightarrow G_x$ is called *regular* if $(\alpha, \lambda) \neq 0$ for all roots α of $(\mathfrak{g}, \mathfrak{g}_x)$. Let us also recall the notation adopted in Lemmas 4.5 and 4.6.

LEMMA 4.9. *Suppose that* $x \in \mathfrak{g}^{rs}$ *and let* $g \in G$ *be such that* $x = \mathrm{Ad}_g(x_{\mathcal{S}})$. *Given any regular coweight* $\lambda : \mathbb{C}^\times \longrightarrow G_x$ *and element* $\widetilde{\mathfrak{b}} \in \mathcal{B}_x$, *one has the following equivalence:*

$$\lim_{t \longrightarrow \infty} \lambda(t) \cdot [g : x_{\mathcal{S}}] = z(\widetilde{\mathfrak{b}}) \Longleftrightarrow \lim_{t \longrightarrow \infty} \lambda(t) \cdot (\gamma_g, (x, x_{\mathcal{S}})) = \Big((e, g^{-1}) \cdot \theta(\widetilde{\mathfrak{b}}), (x, x_{\mathcal{S}})\Big),$$

where the left- (resp. right-) hand side is computed in $\mathrm{Hess}(x, \mathfrak{m})$ *(resp.* $\overline{\mu}_{\mathcal{S}}^{-1}(x)$*).*

PROOF. Let us write $\Pi(\widetilde{\mathfrak{b}}) \subseteq \mathfrak{g}_x^*$ for the set of simple roots determined by $\mathfrak{g}_x$ and $\widetilde{\mathfrak{b}} \in \mathcal{B}_x$. By [**19**, Lemma 7], the G_x-weights of the isotropy representation $T_{z(\widetilde{\mathfrak{b}})}\mathrm{Hess}(x, \mathfrak{m})$ form the set $-\Pi(\widetilde{\mathfrak{b}})$. This has two implications for any regular coweight $\lambda : \mathbb{C}^\times \longrightarrow G_x$. One is that $(\alpha, \lambda) \neq 0$ for all $\widetilde{\mathfrak{b}} \in \mathcal{B}_x$ and G_x-weights α of $T_{z(\widetilde{\mathfrak{b}})}\mathrm{Hess}(x, \mathfrak{m})$. The second implication is the existence of a unique $\widetilde{\mathfrak{b}} \in \mathcal{B}_x$ such that $(\alpha, \lambda) < 0$ for all G_x-weights α of $T_{z(\widetilde{\mathfrak{b}})}\mathrm{Hess}(x, \mathfrak{m})$. These last few sentences and Lemma 4.8 imply the following about a given regular coweight λ and element $\widetilde{\mathfrak{b}} \in \mathcal{B}_x$:

$$\lim_{t \longrightarrow \infty} \lambda(t) \cdot [g : x_{\mathcal{S}}] = z(\widetilde{\mathfrak{b}}) \Longleftrightarrow (\alpha, \lambda) < 0 \text{ for all } \alpha \in -\Pi(\widetilde{\mathfrak{b}}),$$

or equivalently

$$\lim_{t \longrightarrow \infty} \lambda(t) \cdot [g : x_{\mathcal{S}}] = z(\widetilde{\mathfrak{b}}) \Longleftrightarrow (\alpha, \lambda) > 0 \text{ for all } \alpha \in \Pi(\widetilde{\mathfrak{b}}).$$

In light of the previous paragraph, we are reduced to proving that
(4.12)

$$\lim_{t \longrightarrow \infty} \lambda(t) \cdot (\gamma_g, (x, x_{\mathcal{S}})) = \Big((e, g^{-1}) \cdot \theta(\widetilde{\mathfrak{b}}), (x, x_{\mathcal{S}})\Big) \Longleftrightarrow (\alpha, \lambda) > 0 \text{ for all } \alpha \in \Pi(\widetilde{\mathfrak{b}}).$$

Consider the action of $\lambda(t)$ indicated above, noting that it coincides with the action of $(\lambda(t), e) \in G \times G$ on points in $\overline{\mu}_{\mathcal{S}}^{-1}(x) \subseteq T^*\overline{G}(\log D)$. At the same time, the actions of $(\lambda(t), e)$ and (e, g) on $T^*\overline{G}(\log D)$ commute with one another for all $t \in \mathbb{C}^\times$. Acting by (e, g) therefore allows us to reformulate (4.12) as

$$\lim_{t \longrightarrow \infty} \lambda(t) \cdot (\mathfrak{g}_\Delta, (x, x)) = (\theta(\widetilde{\mathfrak{b}}), (x, x)) \Longleftrightarrow (\alpha, \lambda) > 0 \text{ for all } \alpha \in \Pi(\widetilde{\mathfrak{b}}),$$

or equivalently

$$\lim_{t \longrightarrow \infty} (\lambda(t), e) \cdot \mathfrak{g}_\Delta = \theta(\widetilde{\mathfrak{b}}) \Longleftrightarrow (\alpha, \lambda) > 0 \text{ for all } \alpha \in \Pi(\widetilde{\mathfrak{b}}).$$

This last equivalence is a straightforward consequence of Lemma 3.1, and our proof is complete. □

4.3. Some results on gluing. Let us use (3.8) to identify $G \times \mathcal{S}$ with the unique open dense symplectic leaf in $\overline{G \times \mathcal{S}}$. The G-equivariant map $\rho : G \times \mathcal{S} \longrightarrow G \times_B \mathfrak{m}$ in Section 4.1 is then given by

$$\rho(\gamma_g, (\mathrm{Ad}_g(x), x)) = [g : x]$$

for all $(\gamma_g, (\mathrm{Ad}_g(x), x)) \in G \times \mathcal{S} \subseteq \overline{G \times \mathcal{S}}$. One also has the commutative diagram

(4.13)
$$\begin{array}{ccc} G \times \mathcal{S} & \xrightarrow{\quad \rho \quad} & G \times_B \mathfrak{m} \\ & {\scriptstyle \overline{\mu}_{\mathcal{S}}|_{G \times \mathcal{S}}} \searrow \quad \swarrow {\scriptstyle \nu} & \\ & \mathfrak{g} & \end{array} .$$

Now fix $x \in \mathfrak{g}^{\mathrm{rs}}$ and consider the fibres

$$(\overline{\mu}_{\mathcal{S}}\big|_{G\times\mathcal{S}})^{-1}(x) = \overline{\mu}_{\mathcal{S}}^{-1}(x) \cap (G\times\mathcal{S}) =: \overline{\mu}_{\mathcal{S}}^{-1}(x)^{\circ}$$

and

$$\nu^{-1}(x) = \mathrm{Hess}(x,\mathfrak{m}).$$

The commutative diagram (4.13) implies that ρ restricts to a G_x-equivariant morphism

$$\rho_x : \overline{\mu}_{\mathcal{S}}^{-1}(x)^{\circ} \longrightarrow \mathrm{Hess}(x,\mathfrak{m}).$$

LEMMA 4.10. *If $x \in \mathfrak{g}^{rs}$, then there exists a unique G_x-equivariant variety isomorphism*

$$\overline{\rho}_x : \overline{\mu}_{\mathcal{S}}^{-1}(x) \xrightarrow{\cong} \mathrm{Hess}(x,\mathfrak{m})$$

that extends ρ_x.

PROOF. Choose $g\in G$ such that $x = \mathrm{Ad}_g(x_{\mathcal{S}})$. Lemma 4.4 combines with our description of $G\times\mathcal{S}$ as a subset of $\overline{G\times\mathcal{S}}$ to imply that $\overline{\mu}_{\mathcal{S}}^{-1}(x)^{\circ}$ is the unique open dense G_x-orbit in $\overline{\mu}_{\mathcal{S}}^{-1}(x)$. The uniqueness of $\overline{\rho}_x$ is then a consequence of $\overline{\mu}_{\mathcal{S}}^{-1}(x)^{\circ}$ being dense in $\overline{\mu}_{\mathcal{S}}^{-1}(x)$.

To establish existence, recall that $\overline{\mu}_{\mathcal{S}}^{-1}(x)$ and $\mathrm{Hess}(x,\mathfrak{m})$ are smooth, projective, toric G_x-varieties with respective points

$$(\gamma_g, (x, x_{\mathcal{S}})) \quad \text{and} \quad [g : x_{\mathcal{S}}] \tag{4.14}$$

in their open dense G_x-orbits (see Lemmas 4.3 and 4.4). Recall also that elements of $\overline{\mu}_{\mathcal{S}}^{-1}(x)^{G_x}$ and $\mathrm{Hess}(x,\mathfrak{m})^{G_x}$ are in correspondence with elements of $\mathcal{B}_x$ (see Lemmas 4.5 and 4.6), and that the points (4.14) limit to corresponding G_x-fixed points under a given regular coweight (see Lemma 4.9). By these last two sentences, there exists a unique G_x-variety isomorphism

$$\overline{\rho}_x : \overline{\mu}_{\mathcal{S}}^{-1}(x) \xrightarrow{\cong} \mathrm{Hess}(x,\mathfrak{m})$$

satisfying

$$\overline{\rho}_x(\gamma_g, (x, x_{\mathcal{S}})) = [g : x_{\mathcal{S}}].$$

Note that ρ_x is also G_x-equivariant and satisfies

$$\rho_x(\gamma_g, (x, x_{\mathcal{S}})) = [g : x_{\mathcal{S}}].$$

Since $\overline{\rho}_x$ and ρ_x are G_x-equivariant, these last two equations imply that $\overline{\rho}_x$ and ρ_x coincide on

$$G_x \cdot (\gamma_g, (x, x_{\mathcal{S}})) = \overline{\mu}_{\mathcal{S}}^{-1}(x)^{\circ}.$$

□

REMARK 4.11. One may take the following more holistic view of the proof given above. Each of $\overline{\mu}_{\mathcal{S}}^{-1}(\mathfrak{g}^{\mathrm{rs}}) \longrightarrow \mathfrak{g}^{\mathrm{rs}}$ and $\nu^{-1}(\mathfrak{g}^{\mathrm{rs}}) \longrightarrow \mathfrak{g}^{\mathrm{rs}}$ is a family of smooth projective toric varieties. In each family, the fan of the toric variety over $x\in\mathfrak{g}^{\mathrm{rs}}$ is the decomposition of $\mathfrak{g}_x$ into Weyl chambers (see Lemma 4.9 and (4.12); cf. [**19**, Theorem 11] and [**20**, Remark 4.5]). The isomorphisms $\overline{\rho}_x$ simply result from this fibrewise equality of fans. One then anticipates that the following *gluing condition* is satisfied: the isomorphisms $\overline{\rho}_x$ glue together to define a variety isomorphism

$$\overline{\mu}_{\mathcal{S}}^{-1}(\mathfrak{g}^{\mathrm{rs}}) \xrightarrow{\cong} \nu^{-1}(\mathfrak{g}^{\mathrm{rs}}).$$

While this is implicitly taken to be true in the proof of [**5**, Theorem 4.6], we are unable to verify it. If this gluing condition is satisfied, then

$$\begin{array}{ccc} \overline{\mu}_S^{-1}(\mathfrak{g}^{\mathrm{rs}}) & \xrightarrow{\cong} & \nu^{-1}(\mathfrak{g}^{\mathrm{rs}}) \\ & \overline{\mu}_S\big|_{\overline{\mu}_S^{-1}(\mathfrak{g}^{\mathrm{rs}})} \searrow \quad \swarrow \nu\big|_{\nu^{-1}(\mathfrak{g}^{\mathrm{rs}})} & \\ & \mathfrak{g} & \end{array} \tag{4.15}$$

commutes.

4.4. The isomorphism in codimension one. We now undertake a brief digression on transverse intersections in a G-variety. To this end, let X be an arbitrary algebraic variety. Write X_{sing} and $X_{\mathrm{smooth}} := X \setminus X_{\mathrm{sing}}$ for the singular and smooth loci of X, respectively. Given any closed subvariety $Y \subseteq X$, let

$$\mathrm{codim}_X(Y) := \dim X - \dim Y$$

denote the codimension of Y in X.

LEMMA 4.12. *Suppose that X is a smooth G-variety containing a G-invariant closed subvariety $Y \subseteq X$. Let $Z \subseteq X$ be a smooth closed subvariety with the property of being transverse to each G-orbit in X. If W is an irreducible component of $Y \cap Z$, then*

$$\mathrm{codim}_Z(W) \geq \mathrm{codim}_X(Y).$$

PROOF. Set $Y_0 := Y$ and recursively define $Y_{j+1} := (Y_j)_{\mathrm{sing}}$ for all $j \in \mathbb{Z}_{\geq 0}$. One then has a descending chain

$$Y = Y_0 \supseteq Y_1 \supseteq Y_2 \supseteq \cdots$$

of G-invariant closed subvarieties in Y. The locally closed subvarieties $(Y_j)_{\mathrm{smooth}} \subseteq Y$ are also G-invariant, and we observe that these subvarieties have a disjoint union equal to Y.

Let k be the smallest non-negative integer for which $(Y_k)_{\mathrm{smooth}} \cap W \neq \emptyset$. It is then straightforward to verify that $(Y_k)_{\mathrm{smooth}} \cap W = W \setminus Y_{k+1}$. We conclude that $(Y_k)_{\mathrm{smooth}} \cap W$ is an open dense subset of W, and this implies that $W \subseteq Y_k$.

Now choose a point $w \in (Y_k)_{\mathrm{smooth}} \cap W$. Since $(Y_k)_{\mathrm{smooth}}$ is G-invariant and Z is transverse to every G-orbit in X, the varieties $(Y_k)_{\mathrm{smooth}}$ and Z have a transverse intersection at w. It follows that w is a smooth point of $Y_k \cap Z$, and that

$$\dim W \leq \dim(T_w(Y_k \cap Z)) = \dim(T_w Y_k) + \dim Z - \dim X \leq \dim Y + \dim Z - \dim X.$$

This yields the conclusion

$$\mathrm{codim}_Z(W) \geq \mathrm{codim}_X(Y).$$

□

We will ultimately apply this result in a context relevant to our main results. To this end, use (3.6) to identify T^*G with the unique open dense symplectic leaf in $T^*\overline{G}(\log D)$. Consider the open subvariety

$$T^*\overline{G}(\log D)^\circ := T^*G \cup \overline{\mu}_L^{-1}(\mathfrak{g}^{\mathrm{rs}}) \subseteq T^*\overline{G}(\log D)$$

and its complement

$$T^*\overline{G}(\log D)' := T^*\overline{G}(\log D) \setminus T^*\overline{G}(\log D)^\circ,$$

where $\overline{\mu}_L : T^*\overline{G}(\log D) \longrightarrow \mathfrak{g}$ is defined in Section 3.2.

LEMMA 4.13. *We have*

$$\operatorname{codim}_{T^*\overline{G}(\log D)}(T^*\overline{G}(\log D)') \geq 2.$$

PROOF. We begin by observing that

$$T^*\overline{G}(\log D)' = \pi^{-1}(D) \setminus \overline{\mu}_L^{-1}(\mathfrak{g}^{\mathrm{rs}}),$$

where $\pi : T^*\overline{G}(\log D) \longrightarrow \overline{G}$ is the bundle projection and $D = \overline{G} \setminus G$. It therefore suffices to prove that $\overline{\mu}_L^{-1}(\mathfrak{g}^{\mathrm{rs}})$ meets each irreducible component of $\pi^{-1}(D)$.

Now note that

$$D = \bigcup_I \overline{(G \times G)\mathfrak{p}_I \times_{\mathfrak{l}_I} \mathfrak{p}_I^-}$$

is the decomposition of D into irreducible components, where I ranges over all subsets of the form $\Pi \setminus \{\alpha\}$, $\alpha \in \Pi$ [**5**, Section 3.1]. We conclude that

$$\pi^{-1}(D) = \bigcup_I \pi^{-1}\left(\overline{(G \times G)\mathfrak{p}_I \times_{\mathfrak{l}_I} \mathfrak{p}_I^-}\right)$$

is the decomposition of $\pi^{-1}(D)$ into irreducible components. It therefore suffices to prove that $\mu^{-1}(\mathfrak{g}^{\mathrm{rs}})$ meets $\pi^{-1}((G \times G)\mathfrak{p}_I \times_{\mathfrak{l}_I} \mathfrak{p}_I^-)$ for all $I \subseteq \Pi$ of the form $I = \Pi \setminus \{\alpha\}$, $\alpha \in \Pi$.

Choose $x \in \mathfrak{t}^{\mathrm{r}}$ and an element $g \in G$ satisfying $x = \mathrm{Ad}_g(x_{\mathcal{S}})$. We then have

$$(x, x_{\mathcal{S}}) \in (e, g^{-1}) \cdot \mathfrak{p}_I \times_{\mathfrak{l}_I} \mathfrak{p}_I^-$$

for all $I \subseteq \Pi$. We also observe that

$$\left((e, g^{-1}) \cdot \mathfrak{p}_I \times_{\mathfrak{l}_I} \mathfrak{p}_I^-, (x, x_{\mathcal{S}})\right) \in \overline{\mu}_L^{-1}(\mathfrak{g}^{\mathrm{rs}}) \cap \pi^{-1}((G \times G)\mathfrak{p}_I \times_{\mathfrak{l}_I} \mathfrak{p}_I^-)$$

for all $I \subseteq \Pi$. This completes the proof. □

Now consider the open subvariety

$$\overline{G \times \mathcal{S}}^{\circ} := (G \times \mathcal{S}) \cup \overline{\mu}_{\mathcal{S}}^{-1}(\mathfrak{g}^{\mathrm{rs}}) \subseteq \overline{G \times \mathcal{S}}$$

and its complement

$$\overline{G \times \mathcal{S}}' := \overline{G \times \mathcal{S}} \setminus \overline{G \times \mathcal{S}}^{\circ}.$$

LEMMA 4.14. *We have*

$$\operatorname{codim}_{\overline{G \times \mathcal{S}}}(\overline{G \times \mathcal{S}}') \geq 2.$$

PROOF. Our task is to prove that each irreducible component of $\overline{G \times \mathcal{S}}'$ has codimension at least two in $\overline{G \times \mathcal{S}}$. We begin by showing ourselves to be in the situation of Lemma 4.12. To this end, consider $X := T^*\overline{G}(\log D)$ and the action of $G = \{e\} \times G \subseteq G \times G$ on X. Let Y be the subvariety $T^*\overline{G}(\log D)' \subseteq T^*\overline{G}(\log D)$ considered in Lemma 4.13, and set $Z := \overline{G \times \mathcal{S}}$. Since $Z = \overline{\mu}_R^{-1}(\mathcal{S})$, [**17**, Proposition 3.6] implies that Z is transverse to every G-orbit in X. We also note that

$$Y \cap Z = \overline{G \times \mathcal{S}} \setminus (T^*G \cup \overline{\mu}_L^{-1}(\mathfrak{g}^{\mathrm{rs}})) = \overline{G \times \mathcal{S}} \setminus ((G \times \mathcal{S}) \cup \overline{\mu}_{\mathcal{S}}^{-1}(\mathfrak{g}^{\mathrm{rs}})) = \overline{G \times \mathcal{S}}',$$

and that

$$\operatorname{codim}_X(Y) = \operatorname{codim}_{T^*\overline{G}(\log D)}(T^*\overline{G}(\log D)') \geq 2$$

by Lemma 4.13. The desired conclusion now follows immediately from Lemma 4.12. □

Now recall the notation and content of Section 4.1. Consider the open subvariety

$$(G\times_B\mathfrak{m})^\circ := (G\times_B\mathfrak{m}^\times)\cup\nu^{-1}(\mathfrak{g}^{\mathrm{rs}})\subseteq G\times_B\mathfrak{m}$$

and its complement

$$(G\times_B\mathfrak{m})' := G\times_B\mathfrak{m}\setminus(G\times_B\mathfrak{m})^\circ.$$

LEMMA 4.15. *We have*

$$\mathrm{codim}_{G\times_B\mathfrak{m}}((G\times_B\mathfrak{m})')\geq 2.$$

PROOF. It suffices to prove that the open subvariety $\nu^{-1}(\mathfrak{g}^{\mathrm{rs}})\subseteq G\times_B\mathfrak{m}$ meets each irreducible component of $(G\times_B\mathfrak{m})\setminus(G\times_B\mathfrak{m}^\times)$. To this end, suppose that $\beta\in\Pi$ is a simple root and define

$$\begin{aligned}\mathfrak{m}_\beta^\times &:= \mathfrak{b}+\sum_{\alpha\in\Pi\setminus\{\beta\}}(\mathfrak{g}_{-\alpha}\setminus\{0\})\\ &:= \left\{x+\sum_{\alpha\in\Pi\setminus\{\beta\}}c_\alpha e_{-\alpha} : x\in\mathfrak{b}\text{ and } c_\alpha\in\mathbb{C}^\times\text{ for all }\alpha\in\Pi\setminus\{\beta\}\right\}.\end{aligned}$$

The irreducible components of $(G\times_B\mathfrak{m})\setminus(G\times_B\mathfrak{m}^\times)$ are then the subvarieties

$$G\times_B\mathfrak{m}_\beta^\times\subseteq G\times_B\mathfrak{m},\quad \beta\in\Pi.$$

Fix $\beta\in\Pi$ and choose an element $x\in\mathfrak{t}^{\mathrm{r}}$. By [**14**, Lemma 3.1.44], the elements

$$x\quad\text{and}\quad x_\beta := x+\sum_{\alpha\in\Pi\setminus\{\beta\}}e_{-\alpha}$$

are in the same adjoint G-orbit. It follows that $x_\beta\in\mathfrak{g}^{\mathrm{rs}}$, while we also observe that $x_\beta\in\mathfrak{m}_\beta^\times$. These considerations imply that

$$[e:x_\beta]\in(G\times_B\mathfrak{m}_\beta^\times)\cap\nu^{-1}(\mathfrak{g}^{\mathrm{rs}}),$$

forcing

$$(G\times_B\mathfrak{m}_\beta^\times)\cap\nu^{-1}(\mathfrak{g}^{\mathrm{rs}})\neq\emptyset$$

to hold. In light of the previous paragraph, our proof is complete. □

We also require the following elementary lemma in order to prove the main result of this section.

LEMMA 4.16. *Suppose that X and Y are irreducible varieties. Let $U_1,U_2\subseteq X$ and $V_1,V_2\subseteq Y$ be open subsets, and assume that U_1 and U_2 are non-empty. Suppose that $f_1:U_1\longrightarrow V_1$ and $f_2:U_2\longrightarrow V_2$ are variety isomorphisms that agree on points in $U_1\cap U_2$, and let $f:U_1\cup U_2\longrightarrow V_1\cup V_2$ be the result of gluing f_1 and f_2 along $U_1\cap U_2$. Then f is a variety ismomorphism.*

PROOF. Our hypotheses imply that $U_1\cap U_2$ is a non-empty open subset of X. It follows that $f_1(U_1\cap U_2)$ is a non-empty open subset of V_1. One also has $f_1(U_1\cap U_2)\subseteq V_2$, as f_1 and f_2 coincide on $U_1\cap U_2$. We conclude that $f_1(U_1\cap U_2)$ is a non-empty open subset of $V_1\cap V_2$.

Now suppose that $y\in f_1(U_1\cap U_2)$. It follows that

$$f_1(x)=y=f_2(x)$$

for some $x\in U_1\cap U_2$, so that

$$f_1^{-1}(y)=x=f_2^{-1}(y).$$

This establishes that f_1^{-1} and f_2^{-1} coincide on $f_1(U_1 \cap U_2)$. The conclusion of the previous paragraph and the irreducibility of $V_1 \cap V_2$ now imply that f_1^{-1} and f_2^{-1} agree on $V_1 \cap V_2$. We may therefore glue f_1^{-1} and f_2^{-1} along $V_1 \cap V_2$ to obtain a morphism $V_1 \cup V_2 \longrightarrow U_1 \cup U_2$. This morphism is easily checked to be an inverse of f. Our proof is therefore complete. $\square$

The main result of this section is as follows.

THEOREM 4.17. *Assume that the gluing condition in Remark* 4.11 *is satisfied. There exists a unique Hamiltonian G-variety isomorphism*

$$\overline{\rho}^\circ : \overline{G \times \mathcal{S}}^\circ \xrightarrow{\cong} (G \times_B \mathfrak{m})^\circ$$

that extends $\rho : G \times \mathcal{S} \longrightarrow G \times_B \mathfrak{m}^\times$.

PROOF. Uniqueness is an immediate consequence of $G \times \mathcal{S}$ being dense in $\overline{G \times \mathcal{S}}^\circ$. To establish existence, recall the variety isomorphism

$$\overline{\mu}_{\mathcal{S}}^{-1}(\mathfrak{g}^{\mathrm{rs}}) \xrightarrow{\cong} \nu^{-1}(\mathfrak{g}^{\mathrm{rs}}) \tag{4.16}$$

discussed in Remark 4.11. Lemma 4.10 implies that this isomorphism coincides with ρ on the overlap $(G \times \mathcal{S}) \cap \overline{\mu}_{\mathcal{S}}^{-1}(\mathfrak{g}^{\mathrm{rs}})$. We may therefore apply Lemma 4.16 to (4.16) and ρ, where the latter is regarded as an isomorphism onto its image $\rho(G \times \mathcal{S}) = G \times_B \mathfrak{m}^\times$. It follows that ρ extends to a variety isomorphism

$$\overline{G \times \mathcal{S}}^\circ = (G \times \mathcal{S}) \cup \overline{\mu}_{\mathcal{S}}^{-1}(\mathfrak{g}^{\mathrm{rs}}) \xrightarrow{\overline{\rho}^\circ} (G \times_B \mathfrak{m}^\times) \cup \nu^{-1}(\mathfrak{g}^{\mathrm{rs}}) = (G \times_B \mathfrak{m})^\circ.$$

The commutative diagrams (4.13) and (4.15) imply that $\overline{\rho}^\circ$ intertwines the restrictions of $\overline{\mu}_{\mathcal{S}}$ and ν to $\overline{G \times \mathcal{S}}^\circ$ and $(G \times_B \mathfrak{m})^\circ$, respectively.

It remains to establish that $\overline{\rho}^\circ$ is G-equivariant and Poisson. The former is a straightforward consequence of ρ being G-equivariant and $G \times \mathcal{S}$ being dense in $\overline{G \times \mathcal{S}}$. The latter follows from the fact that ρ defines a symplectomorphism between the open dense symplectic leaves $G \times \mathcal{S} \subseteq \overline{G \times \mathcal{S}}$ and $G \times_B \mathfrak{m}^\times \subseteq G \times_B \mathfrak{m}$ (see Proposition 4.2). Our proof is therefore complete. $\square$

REMARK 4.18. In light of Lemmas 4.14 and 4.15, one has the following coarser reformulation of Theorem 4.17: ρ extends to an isomorphism $\overline{G \times \mathcal{S}} \xrightarrow{\cong} G \times_B \mathfrak{m}$ in codimension one. This observation features prominently in proof of Theorem 4.20.

REMARK 4.19. Consider the restricted moment maps

$$\overline{\mu}_{\mathcal{S}}^\circ := \overline{\mu}_{\mathcal{S}}\Big|_{\overline{G \times \mathcal{S}}^\circ} \quad \text{and} \quad \nu^\circ := \nu\Big|_{(G \times_B \mathfrak{m})^\circ},$$

as well as the Hamiltonian G-variety isomorphism $\overline{\rho}^\circ : \overline{G \times \mathcal{S}}^\circ \longrightarrow (G \times_B \mathfrak{m})^\circ$. Theorem 4.17 tells us that

$$\begin{array}{ccc} \overline{G \times \mathcal{S}}^\circ & \xrightarrow{\quad \overline{\rho}^\circ \quad} & (G \times_B \mathfrak{m})^\circ \\ & {\scriptstyle \overline{\mu}_{\mathcal{S}}^\circ} \searrow \quad \swarrow {\scriptstyle \nu^\circ} & \\ & \mathfrak{g} & \end{array} \tag{4.17}$$

commutes, and pulling back along the inclusion $\mathcal{S} \hookrightarrow \mathfrak{g}$ yields a commutative diagram

$$\begin{array}{ccc} \overline{G\times\mathcal{S}}^{\circ}\cap\overline{\mu}_{\mathcal{S}}^{-1}(\mathcal{S})=(\overline{\mu}_{\mathcal{S}}^{\circ})^{-1}(\mathcal{S}) & \longrightarrow & (\nu^{\circ})^{-1}(\mathcal{S})=(G\times_B\mathfrak{m})^{\circ}\cap\nu^{-1}(\mathcal{S}) \\ & \searrow \quad \swarrow & \\ & \mathcal{S} & \end{array}.$$

The horizontal arrow is a Poisson variety isomorphism between the Poisson slices $(\overline{\mu}_{\mathcal{S}}^{\circ})^{-1}(\mathcal{S}) \subseteq \overline{G\times\mathcal{S}}^{\circ}$ and $(\nu^{\circ})^{-1}(\mathcal{S}) \subseteq (G\times_B\mathfrak{m})^{\circ}$, as follows from $\overline{\rho}^{\circ}$ being a Poisson variety isomorphism. On the other hand, [**17**, Example 4.18] explains that $\overline{\mu}_{\mathcal{S}}^{-1}(\mathcal{S}) \longrightarrow \mathcal{S}$ is Balibănu's fibrewise compactified universal centralizer $\overline{\mathcal{Z}_{\mathfrak{g}}} \longrightarrow \mathcal{S}$. Our pullback diagram thereby becomes

$$\begin{array}{ccc} \overline{G\times\mathcal{S}}^{\circ}\cap\overline{\mathcal{Z}_{\mathfrak{g}}} & \xrightarrow{\cong} & (G\times_B\mathfrak{m})^{\circ}\cap\nu^{-1}(\mathcal{S}) \\ & \searrow \quad \swarrow & \\ & \mathcal{S} & \end{array}.$$

The horizontal arrow is precisely the restriction of Balibănu's Poisson isomorphism

$$\overline{\mathcal{Z}_{\mathfrak{g}}} \xrightarrow{\cong} \nu^{-1}(\mathcal{S})$$

to $\overline{G\times\mathcal{S}}^{\circ}\cap\overline{\mathcal{Z}_{\mathfrak{g}}}$, as follows from comparing the proofs of Theorem 4.17 and [**5**, Proposition 4.8].

4.5. The bimeromorphism. While most of our exposition and results have been in the algebraic category, we now adopt some pertinent ideas from complex differential geometry. This will allow us to use the results of [**31**], on which some forthcoming arguments will depend.

Let us briefly recall Remmert's notion of a meromorphic map [**36**], as well as several related concepts. To this end, suppose that X and Y are complex manifolds. We also suppose that $f : X \longrightarrow Y$ is a set-theoretic relation, specified by a subset

$$\mathrm{graph}(f) \subseteq X \times Y.$$

One then calls f a *meromorphic map* if the following conditions are satisfied:

(i) $\mathrm{graph}(f)$ is an analytic subset of $X \times Y$;
(ii) there exists an open dense subset $Z \subseteq X$ for which $f\big|_Z : Z \longrightarrow Y$ is a holomorphic map and
$$\mathrm{graph}(f) = \overline{\mathrm{graph}(f\big|_Z)}$$
in $X \times Y$;
(iii) the projection $X \times Y \longrightarrow X$ restricts to a proper map $\mathrm{graph}(f) \longrightarrow X$.

Suppose that X and Y come equipped with holomorphic G-actions. We then refer to a meromorphic map $f : X \longrightarrow Y$ as being G-equivariant if $\mathrm{graph}(f)$ is invariant under the diagonal G-action on $X \times Y$.

Let $f : X \longrightarrow Y$ and $h : Y \longrightarrow Z$ be meromorphic maps. Declare (f, h) to be composable if there exists an open dense subset $W \subseteq Y$ such that $h\big|_W : W \longrightarrow Z$ is a holomorphic map and

$$f^{-1}(W) := \{x \in X : (x, y) \in \mathrm{graph}(f) \text{ for some } y \in W\}$$

is dense in X. The composite relation

$$h \circ f : X \longrightarrow Z$$

is then a meromorphic map. One calls f a *bimeromorphism* if there exists a meromorphic map $\ell : Y \longrightarrow X$ such that (f, ℓ) and (ℓ, f) are composable and

$$\ell \circ f : X \longrightarrow X \quad \text{and} \quad f \circ \ell : Y \longrightarrow Y$$

are the identity relations.

We may now formulate and prove the main result of this section.

THEOREM 4.20. *Assume that the gluing condition in Remark* 4.11 *is satisfied. There exists a unique bimeromorphism*

$$\overline{\rho} : \overline{G \times \mathcal{S}} \xrightarrow{\cong} G \times_B \mathfrak{m}$$

that extends $\overline{\rho}^{\circ}$. *This bimeromorphism is* G*-equivariant and makes*

$$\begin{array}{ccc} \overline{G \times \mathcal{S}} & \xrightarrow{\quad \overline{\rho} \quad} & G \times_B \mathfrak{m} \\ & {\scriptstyle \overline{\mu}_{\mathcal{S}}} \searrow \quad \swarrow {\scriptstyle \nu} & \\ & \mathfrak{g} & \end{array} \tag{4.18}$$

commute.

PROOF. Uniqueness is an immediate consequence of $G \times \mathcal{S}$ being dense in $\overline{G \times \mathcal{S}}$.

Now recall the G-equivariant closed embedding

$$(\pi, \nu) : G \times_B \mathfrak{m} \longrightarrow G/B \times \mathfrak{g} \tag{4.19}$$

defined in Remark 4.1, and form the G-equivariant composite map

$$\epsilon := (\pi \circ \overline{\rho}^{\circ}, \nu \circ \overline{\rho}^{\circ}) : \overline{G \times \mathcal{S}}^{\circ} \longrightarrow G/B \times \mathfrak{g}.$$

Lemma 4.14 and [**31**, Theorem 3.42] imply that the first component $\pi \circ \overline{\rho}^{\circ}$ extends to a meromorphic map

$$\widetilde{\pi \circ \overline{\rho}^{\circ}} : \overline{G \times \mathcal{S}} \longrightarrow G/B.$$

On the other hand, (4.13) and (4.15) tell us that $\overline{\mu}_{\mathcal{S}}$ extends the second component $\nu \circ \overline{\rho}^{\circ}$ to $\overline{G \times \mathcal{S}}$.

Let us consider the product meromorphic map

$$\widetilde{\epsilon} := (\widetilde{\pi \circ \overline{\rho}^{\circ}}, \overline{\mu}_{\mathcal{S}}) : \overline{G \times \mathcal{S}} \longrightarrow G/B \times \mathfrak{g},$$

i.e. the meromorphic map with graph

$$\begin{aligned} &\mathrm{graph}(\widetilde{\epsilon}) \\ &= \{(\alpha, ([g], x)) \in (\overline{G \times \mathcal{S}}) \times (G/B \times \mathfrak{g}) : (\alpha, [g]) \in \mathrm{graph}(\widetilde{\pi \circ \overline{\rho}^{\circ}}) \text{ and } \overline{\mu}_{\mathcal{S}}(\alpha) = x\}. \end{aligned}$$

The last two sentences of the preceding paragraph imply that $\widetilde{\epsilon}$ extends ϵ. This amounts to having an inclusion

$$\mathrm{graph}(\epsilon) \subseteq \mathrm{graph}(\widetilde{\epsilon}).$$

The irreducibility of $\overline{G \times \mathcal{S}}$ (see [**17**, Theorem 3.14]) and [**10**, Remark 2.3] also force $\mathrm{graph}(\widetilde{\epsilon})$ to be irreducible, and we note that

$$\dim(\mathrm{graph}(\epsilon)) = \dim(\overline{G \times \mathcal{S}}^{\circ}) = \dim(\overline{G \times \mathcal{S}}) = \dim(\mathrm{graph}(\widetilde{\epsilon})).$$

It follows that
$$\mathrm{graph}(\widetilde{\epsilon}) = \overline{\mathrm{graph}(\epsilon)},$$
where the closure is taken in $(\overline{G \times \mathcal{S}}) \times (G/B \times \mathfrak{g})$. This combines with the G-equivariance of ϵ to imply that $\widetilde{\epsilon}$ is G-equivariant. Another consequence is that
$$\mathrm{pr}(\mathrm{graph}(\widetilde{\epsilon})) \subseteq \overline{\mathrm{pr}(\mathrm{graph}(\epsilon))}, \tag{4.20}$$
where
$$\mathrm{pr} : (\overline{G \times \mathcal{S}}) \times (G/B \times \mathfrak{g}) \longrightarrow G/B \times \mathfrak{g}$$
is the obvious projection. We also know that $\mathrm{pr}(\mathrm{graph}(\epsilon))$ is the image of ϵ, and that the latter contains the image of $\rho(G \times \mathcal{S})$ under the closed embedding (4.19). Since $\rho(G \times \mathcal{S})$ is dense in $G \times_B \mathfrak{m}$, this implies that $\overline{\mathrm{pr}(\mathrm{graph}(\epsilon))}$ is the image of (4.19). The inclusion (4.20) thus forces $\mathrm{pr}(\mathrm{graph}(\widetilde{\epsilon}))$ to be contained in the image of (4.19), onto which (4.19) is a G-equivariant isomorphism. One may thereby regard $\widetilde{\epsilon}$ as a G-equivariant meromorphic map
$$\overline{\rho} : \overline{G \times \mathcal{S}} \longrightarrow G \times_B \mathfrak{m}.$$
The fact that $\widetilde{\epsilon}$ extends ϵ then tells us that $\overline{\rho}$ extends $\overline{\rho}^{\circ}$. In particular, $\overline{\rho}$ is a meromorphic extension of ρ.

We now construct an extension of the inverse
$$(\overline{\rho}^{\circ})^{-1} : (G \times_B \mathfrak{m})^{\circ} \longrightarrow \overline{G \times \mathcal{S}}^{\circ}$$
to a meromorphic map $G \times_B \mathfrak{m} \longrightarrow \overline{G \times \mathcal{S}}$. To this end, recall the closed embedding
$$(\pi_{\mathcal{S}}, \overline{\mu}_{\mathcal{S}}) : \overline{G \times \mathcal{S}} \longrightarrow \overline{G} \times \mathfrak{g}$$
discussed in Remark 3.2. One has the composite map
$$\vartheta := (\pi_{\mathcal{S}} \circ (\overline{\rho}^{\circ})^{-1}, \overline{\mu}_{\mathcal{S}} \circ (\overline{\rho}^{\circ})^{-1}) : (G \times_B \mathfrak{m})^{\circ} \longrightarrow \overline{G} \times \mathfrak{g}.$$
By Lemma 4.15 and [**31**, Theorem 3.42], the first component $\pi_{\mathcal{S}} \circ (\overline{\rho}^{\circ})^{-1}$ extends to a meromorphic map
$$\pi_{\mathcal{S}} \widetilde{\circ (\overline{\rho}^{\circ})^{-1}} : G \times_B \mathfrak{m} \longrightarrow \overline{G}.$$
The second component $\overline{\mu}_{\mathcal{S}} \circ (\overline{\rho}^{\circ})^{-1}$ extends to ν on $G \times_B \mathfrak{m}$, as follows from our earlier observation that $\overline{\mu}_{\mathcal{S}}$ extends $\nu \circ \overline{\rho}^{\circ}$.

Now consider the product meromorphic map
$$\widetilde{\vartheta} := (\pi_{\mathcal{S}} \widetilde{\circ (\overline{\rho}^{\circ})^{-1}}, \nu) : G \times_B \mathfrak{m} \longrightarrow \overline{G} \times \mathfrak{g},$$
i.e. the meromorphic map with graph
$$\begin{aligned}&\mathrm{graph}(\widetilde{\vartheta})\\ &\quad = \{(\alpha, (\gamma, x)) \in (G \times_B \mathfrak{m}) \times (\overline{G} \times \mathfrak{g}) : (\alpha, \gamma) \in \mathrm{graph}(\pi_{\mathcal{S}} \widetilde{\circ (\overline{\rho}^{\circ})^{-1}}) \text{ and } \nu(\alpha) = x\}.\end{aligned}$$
One can then use direct analogues of arguments made about $\widetilde{\epsilon}$ to establish the following: $\widetilde{\vartheta}$ extends ϑ, and it may be regarded as a meromorphic map
$$\overline{\rho^{-1}} : G \times_B \mathfrak{m} \longrightarrow \overline{G \times \mathcal{S}}$$
that extends $(\overline{\rho}^{\circ})^{-1}$ and ρ^{-1}. Since $\overline{\rho}$ extends ρ, one deduces that $\overline{\rho}$ and $\overline{\rho^{-1}}$ are inverses.

It remains only to establish that (4.18) commutes. We begin by interpreting commutativity in (4.13) as the statement that
$$\mathrm{graph}(\nu \circ \rho) = \mathrm{graph}(\overline{\mu}_{\mathcal{S}}\big|_{G \times \mathcal{S}}).$$

On the other hand, it is straightforward to establish that

$$\mathrm{graph}(\nu\circ\overline{\rho}) = \overline{\mathrm{graph}(\nu\circ\rho)} \subseteq (\overline{G\times\mathcal{S}})\times\mathfrak{g} \quad \text{and}$$
$$\mathrm{graph}(\overline{\mu}_{\mathcal{S}}) = \overline{\mathrm{graph}(\overline{\mu}_{\mathcal{S}}\big|_{G\times\mathcal{S}})} \subseteq (\overline{G\times\mathcal{S}})\times\mathfrak{g}.$$

These last two sentences imply that

$$\mathrm{graph}(\nu\circ\overline{\rho}) = \mathrm{graph}(\overline{\mu}_{\mathcal{S}}),$$

i.e. (4.18) commutes. This completes the proof. □

4.6. The fibre of $\overline{\mu}_{\mathcal{S}}$ over 0. One is naturally motivated to improve Theorem 4.20. Section 4.7 represents a step in this direction, and it uses the next few results. To this end, recall notation, conventions, and facts discussed in Sections 2.2 and 3.1.

PROPOSITION 4.21. *Suppose that $\gamma\in\overline{G}$. One then has*

$$(0,\xi)\in\gamma \iff \gamma\in(G\times B^-)\mathfrak{b}\times_{\mathfrak{t}}\mathfrak{b}^-.$$

PROOF. To prove the forward implication, we assume that $(0,\xi)\in\gamma$. One may write

$$\gamma = (g_1,g_2)\cdot\mathfrak{p}_I\times_{\mathfrak{l}_I}\mathfrak{p}_I^-$$

for suitable elements $(g_1,g_2)\in G\times G$ and a subset $I\subseteq\Pi$. It follows that

$$(0,\mathrm{Ad}_{g_2^{-1}}(\xi))\in\mathfrak{p}_I\times_{\mathfrak{l}_I}\mathfrak{p}_I^-,$$

or equivalently that

$$\mathrm{Ad}_{g_2^{-1}}(\xi)\in\mathfrak{u}_I^-.$$

This combines with [**28**, Theorem 5.3] and the fact that $\mathrm{Ad}_{g_2^{-1}}(\xi)\in\mathfrak{g}^{\mathrm{r}}$ to imply that $I=\emptyset$. We deduce that

$$\gamma = (g_1,g_2)\cdot\mathfrak{b}\times_{\mathfrak{t}}\mathfrak{b}^- \quad \text{and} \quad \xi\in\mathrm{Ad}_{g_2}(\mathfrak{b}^-).$$

On the other hand, $\mathfrak{b}^-$ is the unique Borel subalgebra of $\mathfrak{g}$ that contains ξ. This implies that

$$\mathrm{Ad}_{g_2}(\mathfrak{b}^-) = \mathfrak{b}^-,$$

i.e. $g_2\in B^-$. We therefore have

$$\gamma = (g_1,g_2)\cdot\mathfrak{b}\times_{\mathfrak{t}}\mathfrak{b}^- \in (G\times B^-)\mathfrak{b}\times_{\mathfrak{t}}\mathfrak{b}^-.$$

Now assume that $\gamma\in(G\times B^-)\mathfrak{b}\times_{\mathfrak{t}}\mathfrak{b}^-$, i.e. that

$$\gamma = (g,b)\cdot\mathfrak{b}\times_{\mathfrak{t}}\mathfrak{b}^-$$

for some $g\in G$ and $b\in B^-$. Since

$$\mathrm{Ad}_{b^{-1}}(\xi)\in\mathfrak{u}^-,$$

we must have

$$(0,\mathrm{Ad}_{b^{-1}}(\xi))\in\mathfrak{b}\times_{\mathfrak{t}}\mathfrak{b}^-.$$

This amounts to the statement that

$$(0,\xi)\in(g,b)\cdot\mathfrak{b}\times_{\mathfrak{t}}\mathfrak{b}^- = \gamma,$$

completing the proof. □

COROLLARY 4.22. *There is an isomorphism of varieties*

$$\overline{\mu}_{\mathcal{S}}^{-1}(0)\cong G/B.$$

PROOF. Note that $x_{\mathcal{S}} = \xi$ if $x = 0$. It follows that

$$\begin{aligned}\overline{\mu}_{\mathcal{S}}^{-1}(0) &= \{(\gamma,(0,\xi)) : \gamma \in \overline{G} \text{ and } (0,\xi) \in \gamma\}\\ &\cong \{\gamma \in \overline{G} : (0,\xi) \in \gamma\}\\ &= (G \times B^-)\mathfrak{b} \times_{\mathfrak{t}} \mathfrak{b}^-,\end{aligned}$$

where the last line comes from Proposition 4.21. At the same time, note that the $(G \times G)$-stabilizer of $\mathfrak{b} \times_{\mathfrak{t}} \mathfrak{b}^- \in \overline{G}$ is $B \times B^-$ (see (4.11)). We therefore have

$$\overline{\mu}_{\mathcal{S}}^{-1}(0) \cong (G \times B^-)/(B \times B^-) \cong G/B.$$

□

REMARK 4.23. Observe that $\nu^{-1}(0)$ is precisely the zero-section of the vector bundle $G \times_B \mathfrak{m} \longrightarrow G/B$. It follows that $\nu^{-1}(0) \cong G/B$, or equivalently

$$\nu^{-1}(0) \cong \overline{\mu}_{\mathcal{S}}^{-1}(0).$$

4.7. The case $\mathfrak{g} = \mathfrak{sl}_2$. We now specialize some of our constructions to the case $\mathfrak{g} = \mathfrak{sl}_2$. This culminates in Theorem 4.25, to which the following lemma is relevant.

LEMMA 4.24. *If $\mathfrak{g} = \mathfrak{sl}_2$, then one has a variety isomorphism*

$$\overline{\mu}_{\mathcal{S}}^{-1}(x) \cong \mathbb{P}^1$$

for each $x \in \mathfrak{g}$.

PROOF. The case $x = 0$ follows from Corollary 4.22 and the fact that $G/B \cong \mathbb{P}^1$. We may therefore assume that $x \neq 0$. Since $\mathfrak{g} = \mathfrak{sl}_2$, this is equivalent to having $x \in \mathfrak{g}^{\mathrm{r}}$. Lemma 4.4 and the description

$$\overline{\mu}_{\mathcal{S}}^{-1}(x) = \{(\gamma,(x,x_{\mathcal{S}})) : \gamma \in \overline{G} \text{ and } (x,x_{\mathcal{S}}) \in \gamma\}$$

now imply the following: $\overline{\mu}_{\mathcal{S}}^{-1}(x)$ is isomorphic to the closure of a free orbit of $G_x = G_x \times \{e\} \subseteq G \times G$ in $\overline{G}$. By [**6**, Theorem 5.2], the latter is isomorphic to the closure of a free G_x-orbit in $G/B \cong \mathbb{P}^1$. Dimension considerations force this new orbit closure to be $\mathbb{P}^1$ itself. The last three sentences imply that

$$\overline{\mu}_{\mathcal{S}}^{-1}(x) \cong \mathbb{P}^1.$$

□

This facilitates the following improvement to Theorem 4.20.

THEOREM 4.25. *Assume that the gluing condition in Remark* 4.11 *is satisfied. If $\mathfrak{g} = \mathfrak{sl}_2$, then the bimeromorphism $\overline{\rho} : \overline{G \times \mathcal{S}} \to G \times_B \mathfrak{m}$ is a biholomorphism.*

PROOF. In light of Theorem 4.20, we must prove that

$$\overline{\rho}^{\circ} : \overline{G \times \mathcal{S}}^{\circ} \xrightarrow{\cong} (G \times_B \mathfrak{m})^{\circ}$$

extends to a biholomorphism

$$\overline{G \times \mathcal{S}} \xrightarrow{\cong} G \times_B \mathfrak{m}.$$

We begin by observing that $\mathfrak{m} = \mathfrak{g}$, as $\mathfrak{g} = \mathfrak{sl}_2$. The closed embedding (4.19) is therefore an isomorphism

$$(\pi, \nu) : G \times_B \mathfrak{m} \xrightarrow{\cong} G/B \times \mathfrak{g}.$$

It thus suffices to prove that

$$\varphi := (\pi, \nu) \circ \overline{\rho}^{\circ} : \overline{G \times \mathcal{S}}^{\circ} \longrightarrow G/B \times \mathfrak{g}$$

extends to a biholomorphism

$$\overline{G \times \mathcal{S}} \xrightarrow{\cong} G/B \times \mathfrak{g}.$$

Note that the biholomorphism group of $G/B \times \mathfrak{g}$ acts transitively, and that $G/B \times \mathfrak{g}$ is an unbranched cover of the compact Kähler manifold $\mathbb{P}^1 \times (\mathbb{C}^3/\mathbb{Z}^6)$. We also know φ to be a biholomorphism onto its image, and that $\overline{G \times \mathcal{S}}^{\circ}$ has a complement of codimension at least two in $\overline{G \times \mathcal{S}}$ (see Lemma 4.14). These last two sentences allow us to apply [**31**, Theorem 3.48] and deduce that φ extends to a local biholomorphism

$$\widetilde{\varphi} : \overline{G \times \mathcal{S}} \longrightarrow G/B \times \mathfrak{g}.$$

The commutative diagrams (4.2) and (4.17) are also easily seen to imply that

$$\begin{array}{ccc} \overline{G \times \mathcal{S}} & \xrightarrow{\quad \widetilde{\varphi} \quad} & G/B \times \mathfrak{g} \\ & {\scriptstyle \overline{\mu}_{\mathcal{S}}} \searrow \quad \swarrow {\scriptstyle \mathrm{pr}_{\mathfrak{g}}} & \\ & \mathfrak{g} & \end{array}$$

commutes, where $\mathrm{pr}_{\mathfrak{g}} : G/B \times \mathfrak{g} \longrightarrow \mathfrak{g}$ is projection to $\mathfrak{g}$. It follows that the local biholomorphism $\widetilde{\varphi}$ restricts to a locally injective holomorphic map

$$\widetilde{\varphi}_x : \overline{\mu}_{\mathcal{S}}^{-1}(x) \longrightarrow \mathrm{pr}_{\mathfrak{g}}^{-1}(x)$$

for each $x \in \mathfrak{g}$. The domain and codomain of $\widetilde{\varphi}_x$ are isomorphic to $\mathbb{P}^1$ (see Lemma 4.24), and a locally injective holomorphic map $\mathbb{P}^1 \longrightarrow \mathbb{P}^1$ is a biholomorphism. We conclude that $\widetilde{\varphi}_x$ is a biholomorphism for each $x \in \mathfrak{g}$. Our local biholomorphism $\widetilde{\varphi}$ is therefore bijective, so that it must be a biholomorphism. This completes the proof. □

4.8. A conjecture and its implications for Hessenberg varieties. Theorems 4.20 and 4.25 and Remark 4.23 lend plausibility to the following conjecture.

CONJECTURE 4.26. The bimeromorphism $\overline{\rho} : \overline{G \times \mathcal{S}} \xrightarrow{\cong} G \times_B \mathfrak{m}$ in Theorem 4.20 is a biholomorphism.

We now formulate some implications of this conjecture for the geometry of Hessenberg varieties. To this end, assume that Conjecture 4.26 is true. Consider the bundle projection $\pi : T^*\overline{G}(\log D) \longrightarrow \overline{G}$ and the composite map

$$\psi := \pi\Big|_{\overline{G \times \mathcal{S}}} \circ \overline{\rho}^{-1} : G \times_B \mathfrak{m} \longrightarrow \overline{G}.$$

Let us write

$$\psi_x := \psi\big|_{\mathrm{Hess}(x,\mathfrak{m})} : \mathrm{Hess}(x, \mathfrak{m}) \longrightarrow \overline{G}$$

for the restriction of ψ to a Hessenberg variety $\mathrm{Hess}(x, \mathfrak{m}) \subseteq G \times_B \mathfrak{m}$.

COROLLARY 4.27. *Assume that Conjecture 4.26 is true. If $x \in \mathfrak{g}$, then*

$$\psi_x : \mathrm{Hess}(x, \mathfrak{m}) \longrightarrow \overline{G}$$

is a closed embedding of algebraic varieties. This embedding is G_x-equivariant with respect to the action of $G_x = G_x \times \{e\} \subseteq G \times G$ on $\overline{G}$.

PROOF. Theorem 4.20 and Conjecture 4.26 tell us that $\overline{\rho}$ restricts to a biholomorphism

$$\overline{\rho}_x : \overline{\mu}_{\mathcal{S}}^{-1}(x) \xrightarrow{\cong} \nu^{-1}(x) = \mathrm{Hess}(x, \mathfrak{m})$$

of complex analytic spaces. Since $\overline{\mu}_{\mathcal{S}}^{-1}(x)$ and $\mathrm{Hess}(x, \mathfrak{m})$ are projective (see Remarks 3.2 and 4.1), it follows that $\overline{\rho}_x$ is a variety isomorphism (e.g. by [**11**, Corollary A.4]). At the same time, Remark 3.2 implies that π restricts to a closed embedding

$$\pi_x : \overline{\mu}_{\mathcal{S}}^{-1}(x) \longrightarrow \overline{G}, \quad (\gamma, (x, x_{\mathcal{S}})) \mapsto \gamma, \quad (\gamma, (x, x_{\mathcal{S}})) \in \overline{\mu}_{\mathcal{S}}^{-1}(x).$$

The composite map

$$\psi_x = \pi_x \circ (\overline{\rho}_x)^{-1} : \mathrm{Hess}(x, \mathfrak{m}) \longrightarrow \overline{G}$$

is therefore a closed embedding of algebraic varieties. The claim about equivariance follows from the G-equivariance of $\overline{\rho}$, together with the fact that π is $(G \times G)$-equivariant. □

One can be reasonably explicit about the image of ψ_x, especially if $x \in \mathfrak{g}^{\mathrm{r}}$. To elaborate on this, recall the $(G \times G)$-equivariant locally closed embedding

$$\kappa : G \longrightarrow \overline{G} \subseteq \mathrm{Gr}(n, \mathfrak{g} \oplus \mathfrak{g})$$

defined in (3.1). Given any $x \in \mathfrak{g}$, let us write $\overline{G_x}$ for the closure of $\kappa(G_x)$ in $\overline{G}$.

PROPOSITION 4.28. *Assume that Conjecture 4.26 is true. If $x \in \mathfrak{g}$, then the following statements hold.*

(i) *We have*

$$\mathrm{image}(\psi_x) = \{\gamma \in \overline{G} : (x, x_{\mathcal{S}}) \in \gamma\}.$$

(ii) *If $x \in \mathfrak{g}^{\mathrm{r}}$, then*

$$\mathrm{image}(\psi_x) = (e, g^{-1}) \cdot \overline{G_x}$$

for any $g \in G$ satisfying $x = \mathrm{Ad}_g(x_{\mathcal{S}})$.

PROOF. To prove (i), recall the notation used in the proof of Corollary 4.27. The image of ψ_x coincides with that of π_x, and the latter is

$$\{\gamma \in \overline{G} : (x, x_{\mathcal{S}}) \in \gamma\}.$$

We now verify (ii). Note that

$$(x, x_{\mathcal{S}}) = (\mathrm{Ad}_{hg}(x_{\mathcal{S}}), x_{\mathcal{S}}) \in \gamma_{hg}$$

for all $h \in G_x$. This combines with (i) to imply that $\gamma_{hg} \in \mathrm{image}(\psi_x)$ for all $h \in G_x$, i.e.

$$\kappa(G_x g) \subseteq \mathrm{image}(\psi_x).$$

Since $\mathrm{image}(\psi_x)$ is closed in $\overline{G}$, it must therefore contain the closure $\overline{\kappa(G_x g)}$ of $\kappa(G_x g)$ in $\overline{G}$. We also note that $\mathrm{image}(\psi_x) \cong \mathrm{Hess}(x, \mathfrak{m})$ is ℓ-dimensional and irreducible (see [**33**, Corollaries 3 and 14]), and that

$$\dim(\overline{\kappa(G_x g)}) = \dim(G_x) = \ell.$$

It follows that

$$\overline{\kappa(G_x g)} = \mathrm{image}(\psi_x).$$

It remains only to invoke the $(G \times G)$-equivariance of κ and conclude that

$$\overline{\kappa(G_x g)} = (e, g^{-1}) \cdot \overline{\kappa(G_x)} = (e, g^{-1}) \cdot \overline{G_x}.$$

□

REMARK 4.29. Assume that Conjecture 4.26 is true. If $x \in \mathcal{S}$, then one may apply Proposition 4.28(ii) with $g = e$ and conclude that $\mathrm{image}(\psi_x) = \overline{G_x}$. Corollary 4.27 therefore yields a G_x-equivariant isomorphism

$$\mathrm{Hess}(x, \mathfrak{m}) \xrightarrow{\cong} \overline{G_x}.$$

This is precisely the isomorphism obtained in [**5**, Corollary 4.10].

Notation

- $\mathfrak{g}$ — finite-dimensional complex semisimple Lie algebra
- n — dimension of $\mathfrak{g}$
- ℓ — rank of $\mathfrak{g}$
- $\mathfrak{g}^{\mathrm{r}}$ — subset of regular elements in $\mathfrak{g}$
- V^{r} — the intersection $V \cap \mathfrak{g}^{\mathrm{r}}$ for any subset $V \subseteq \mathfrak{g}$
- $\mathfrak{g}^{\mathrm{rs}}$ — subset of regular semisimple elements in $\mathfrak{g}$
- $\langle \cdot, \cdot \rangle$ — Killing form on $\mathfrak{g}$
- $V^{\perp}$ — annihilator of $V \subseteq \mathfrak{g}$ with respect to $\langle \cdot, \cdot \rangle$
- G — adjoint group of $\mathfrak{g}$
- G_x — G-stabilizer of $x \in \mathfrak{g}$ with respect to the adjoint action
- $\mu = (\mu_L, \mu_R) : T^*G \longrightarrow \mathfrak{g} \oplus \mathfrak{g}$ — moment map for the $(G \times G)$-action on T^*G
- $\mu_{\mathcal{S}} : G \times \mathcal{S} \longrightarrow \mathfrak{g}$ — restriction of μ_L to $G \times \mathcal{S}$
- τ — fixed principal $\mathfrak{sl}_2$-triple in $\mathfrak{g}$
- $\mathcal{S}$ — Slodowy slice associated to τ
- $\chi : \mathfrak{g} \longrightarrow \mathrm{Spec}(\mathbb{C}[\mathfrak{g}]^G)$ — adjoint quotient
- $x_{\mathcal{S}}$ — unique point at which $\mathcal{S}$ meets the fibre $\chi^{-1}(\chi(x))$
- $\overline{G}$ — De Concini–Procesi wonderful compactification of G
- $\mathfrak{g}_{\Delta}$ — diagonal in $\mathfrak{g} \oplus \mathfrak{g}$
- γ_g — the point $(g, e) \cdot \mathfrak{g}_{\Delta} \in \overline{G}$
- $\kappa : G \longrightarrow \overline{G}$ — open embedding defined by $g \mapsto \gamma_g$
- D — the divisor $\overline{G} \setminus \kappa(G)$
- $\overline{G_x}$ — closure of $\kappa(G_x)$ in $\overline{G}$
- $T^*\overline{G}(\log(D))$ — log cotangent bundle of $(\overline{G}, D)$
- $\overline{\mu} = (\overline{\mu}_L, \overline{\mu}_R) : T^*\overline{G}(\log(D)) \longrightarrow \mathfrak{g} \oplus \mathfrak{g}$ — moment map for the $(G \times G)$-action on $T^*\overline{G}(\log(D))$
- $\overline{G \times \mathcal{S}}$ — the Poisson slice $\overline{\mu}_R^{-1}(\mathcal{S})$
- $\overline{\mu}_{\mathcal{S}} : \overline{G \times \mathcal{S}} \longrightarrow \mathfrak{g}$ — restriction of $\overline{\mu}_L$ to $\overline{G \times \mathcal{S}}$
- $\mathcal{Z}_{\mathfrak{g}}$ — universal centralizer of $\mathfrak{g}$
- $\overline{\mathcal{Z}_{\mathfrak{g}}}$ — Bălibanu's fibrewise compactification of $\mathcal{Z}_{\mathfrak{g}}$
- $\mathcal{B}$ — abstract flag variety of all Borel subalgebras in $\mathfrak{g}$
- $\mathfrak{m}$ — standard Hessenberg subspace
- $\nu : G \times_B \mathfrak{m} \longrightarrow \mathfrak{g}$ — moment map for the G-action on $G \times_B \mathfrak{m}$
- $\mathrm{Hess}(x, \mathfrak{m})$ — fibre of ν over $x \in \mathfrak{g}$
- Π — set of simple roots
- $\mathfrak{p}_I, \mathfrak{p}_I^-$ — standard parabolic, opposite parabolic associated to $I \subseteq \Pi$
- $\mathfrak{l}_I$ — standard Levi subalgebra $\mathfrak{p}_I \cap \mathfrak{p}_I^-$ associated with $I \subseteq \Pi$

Acknowledgments

We gratefully acknowledge Ana Bălibanu for constructive conversations and suggestions, as well as the referee for a careful reading and pertinent comments.

References

[1] Hiraku Abe and Peter Crooks, *Hessenberg varieties for the minimal nilpotent orbit*, Pure Appl. Math. Q. **12** (2016), no. 2, 183–223, DOI 10.4310/PAMQ.2016.v12.n2.a1. MR3767215

[2] Hiraku Abe and Peter Crooks, *Hessenberg varieties, Slodowy slices, and integrable systems*, Math. Z. **291** (2019), no. 3-4, 1093–1132, DOI 10.1007/s00209-019-02235-7. MR3936100

[3] Hiraku Abe, Lauren DeDieu, Federico Galetto, and Megumi Harada, *Geometry of Hessenberg varieties with applications to Newton-Okounkov bodies*, Selecta Math. (N.S.) **24** (2018), no. 3, 2129–2163, DOI 10.1007/s00029-018-0405-3. MR3816501

[4] Hiraku Abe, Tatsuya Horiguchi, and Mikiya Masuda, *The cohomology rings of regular semisimple Hessenberg varieties for* $h = (h(1), n, \ldots, n)$, J. Comb. **10** (2019), no. 1, 27–59, DOI 10.4310/joc.2019.v10.n1.a2. MR3890915

[5] A. Bălibanu, *The partial compactification of the universal centralizer*, `arXiv:1710.06327`, (2021), 28pp.

[6] Ana Bălibanu, *The Peterson variety and the wonderful compactification*, Represent. Theory **21** (2017), 132–150, DOI 10.1090/ert/499. MR3673527

[7] A. Bălibanu and P. Crooks, *Perverse sheaves and the cohomology of regular Hessenberg varieties*, To appear in Transform. Groups. `arXiv:2004.07970`, (2020), 24pp.

[8] Roger Bielawski, *Hyper-Kähler structures and group actions*, J. London Math. Soc. (2) **55** (1997), no. 2, 400–414, DOI 10.1112/S0024610796004723. MR1438643

[9] Roger Bielawski, *Slices to sums of adjoint orbits, the Atiyah-Hitchin manifold, and Hilbert schemes of points*, Complex Manifolds **4** (2017), no. 1, 16–36, DOI 10.1515/coma-2017-0003. MR3610019

[10] L. Biliotti and A. Ghigi, *Meromorphic limits of automorphisms*, Transform. Groups **26** (2021), no. 4, 1147–1168, DOI 10.1007/s00031-020-09551-x. MR4372941

[11] Christina Birkenhake and Herbert Lange, *Complex abelian varieties*, 2nd ed., Grundlehren der mathematischen Wissenschaften [Fundamental Principles of Mathematical Sciences], vol. 302, Springer-Verlag, Berlin, 2004, DOI 10.1007/978-3-662-06307-1. MR2062673

[12] Michel Brion, *Vanishing theorems for Dolbeault cohomology of log homogeneous varieties*, Tohoku Math. J. (2) **61** (2009), no. 3, 365–392, DOI 10.2748/tmj/1255700200. MR2568260

[13] Patrick Brosnan and Timothy Y. Chow, *Unit interval orders and the dot action on the cohomology of regular semisimple Hessenberg varieties*, Adv. Math. **329** (2018), 955–1001, DOI 10.1016/j.aim.2018.02.020. MR3783432

[14] Neil Chriss and Victor Ginzburg, *Representation theory and complex geometry*, Modern Birkhäuser Classics, Birkhäuser Boston, Ltd., Boston, MA, 2010. Reprint of the 1997 edition, DOI 10.1007/978-0-8176-4938-8. MR2838836

[15] Peter Crooks, *An equivariant description of certain holomorphic symplectic varieties*, Bull. Aust. Math. Soc. **97** (2018), no. 2, 207–214, DOI 10.1017/S0004972717001095. MR3772649

[16] Peter Crooks and Steven Rayan, *Abstract integrable systems on hyperkähler manifolds arising from Slodowy slices*, Math. Res. Lett. **26** (2019), no. 1, 9–33, DOI 10.4310/MRL.2019.v26.n1.a2. MR3963973

[17] Peter Crooks and Markus Röser, *The log symplectic geometry of Poisson slices*, J. Symplectic Geom. **20** (2022), no. 1, 135–189. MR4518250

[18] Peter Crooks and Maarten van Pruijssen, *An application of spherical geometry to hyperkähler slices*, Canad. J. Math. **73** (2021), no. 3, 687–716, DOI 10.4153/S0008414X20000127. MR4282013

[19] F. De Mari, C. Procesi, and M. A. Shayman, *Hessenberg varieties*, Trans. Amer. Math. Soc. **332** (1992), no. 2, 529–534, DOI 10.2307/2154181. MR1043857

[20] S. Evens, and B. F. Jones, *On the wonderful compactification*, `arXiv:0801.0456`, (2008), 28pp.

[21] Pedro Frejlich and Ioan Mărcuţ, *Normal forms for Poisson maps and symplectic groupoids around Poisson transversals*, Lett. Math. Phys. **108** (2018), no. 3, 711–735, DOI 10.1007/s11005-017-1007-2. MR3765976

[22] Ryushi Goto, *Rozansky-Witten invariants of log symplectic manifolds*, Integrable systems, topology, and physics (Tokyo, 2000), Contemp. Math., vol. 309, Amer. Math. Soc., Providence, RI, 2002, pp. 69–84, DOI 10.1090/conm/309/05342. MR1953353

[23] Marco Gualtieri, Songhao Li, Álvaro Pelayo, and Tudor S. Ratiu, *The tropical momentum map: a classification of toric log symplectic manifolds*, Math. Ann. **367** (2017), no. 3-4, 1217–1258, DOI 10.1007/s00208-016-1427-9. MR3623224

[24] Marco Gualtieri and Brent Pym, *Poisson modules and degeneracy loci*, Proc. Lond. Math. Soc. (3) **107** (2013), no. 3, 627–654, DOI 10.1112/plms/pds090. MR3100779

[25] Victor Guillemin, Eva Miranda, and Ana Rita Pires, *Symplectic and Poisson geometry on b-manifolds*, Adv. Math. **264** (2014), 864–896, DOI 10.1016/j.aim.2014.07.032. MR3250302

[26] Victor W. Guillemin, Eva Miranda, and Jonathan Weitsman, *On geometric quantization of b-symplectic manifolds*, Adv. Math. **331** (2018), 941–951, DOI 10.1016/j.aim.2018.04.003. MR3804693

[27] Megumi Harada and Martha E. Precup, *The cohomology of abelian Hessenberg varieties and the Stanley-Stembridge conjecture*, Algebr. Comb. **2** (2019), no. 6, 1059–1108, DOI 10.5802/alco.76. Revised edition of [MR3940624]. MR4049838

[28] Bertram Kostant, *The principal three-dimensional subgroup and the Betti numbers of a complex simple Lie group*, Amer. J. Math. **81** (1959), 973–1032, DOI 10.2307/2372999. MR114875

[29] Bertram Kostant, *Lie group representations on polynomial rings*, Amer. J. Math. **85** (1963), 327–404, DOI 10.2307/2373130. MR158024

[30] Bertram Kostant, *Fomenko-Mischenko theory, Hessenberg varieties, and polarizations*, Lett. Math. Phys. **90** (2009), no. 1-3, 253–285, DOI 10.1007/s11005-009-0344-1. MR2565041

[31] Benjamin McKay, *Extension phenomena for holomorphic geometric structures*, SIGMA Symmetry Integrability Geom. Methods Appl. **5** (2009), Paper 058, 45, DOI 10.3842/SIGMA.2009.058. MR2529189

[32] Gregory W. Moore and Yuji Tachikawa, *On 2d TQFTs whose values are holomorphic symplectic varieties*, String-Math 2011, Proc. Sympos. Pure Math., vol. 85, Amer. Math. Soc., Providence, RI, 2012, pp. 191–207, DOI 10.1090/pspum/085/1379. MR2985331

[33] Martha Precup, *The Betti numbers of regular Hessenberg varieties are palindromic*, Transform. Groups **23** (2018), no. 2, 491–499, DOI 10.1007/s00031-017-9442-9. MR3805214

[34] Brent Pym and Travis Schedler, *Holonomic Poisson manifolds and deformations of elliptic algebras*, Geometry and physics. Vol. II, Oxford Univ. Press, Oxford, 2018, pp. 681–703. MR3931792

[35] Olga Radko, *A classification of topologically stable Poisson structures on a compact oriented surface*, J. Symplectic Geom. **1** (2002), no. 3, 523–542. MR1959058

[36] Reinhold Remmert, *Holomorphe und meromorphe Abbildungen komplexer Räume* (German), Math. Ann. **133** (1957), 328–370, DOI 10.1007/BF01342886. MR92996

[37] John Shareshian and Michelle L. Wachs, *Chromatic quasisymmetric functions*, Adv. Math. **295** (2016), 497–551, DOI 10.1016/j.aim.2015.12.018. MR3488041

[38] Julianna S. Tymoczko, *Paving Hessenberg varieties by affines*, Selecta Math. (N.S.) **13** (2007), no. 2, 353–367, DOI 10.1007/s00029-007-0038-4. MR2361098

[39] Julianna S. Tymoczko, *Permutation actions on equivariant cohomology of flag varieties*, Toric topology, Contemp. Math., vol. 460, Amer. Math. Soc., Providence, RI, 2008, pp. 365–384, DOI 10.1090/conm/460/09030. MR2428368

Department of Mathematics and Statistics, Utah State University, 3900 Old Main Hill, Logan, Utah 84322

Email address: peter.crooks@usu.edu

Fachbereich Mathematik, Universität Hamburg, Bundesstrasse 55 (Geomatikum), 20146 Hamburg, Germany

Email address: markus.roeser@uni-hamburg.de

Contemporary Mathematics
Volume **790**, 2023
https://doi.org/10.1090/conm/790/15858

Geometry of logarithmic derivations of hyperplane arrangements

Graham Denham and Avi Steiner

Abstract. We study the Hadamard product of the linear forms defining a hyperplane arrangement with those of its dual, which we view as generating an ideal in a certain polynomial ring. We use this ideal, which we call the ideal of pairs, to study logarithmic derivations and critical set varieties of arrangements in a way which is symmetric with respect to matroid duality. Our main result exhibits the variety of the ideal of pairs as a subspace arrangement whose components correspond to cyclic flats of the arrangement. As a corollary, we are able to give geometric explanations of some freeness and projective dimension results due to Ziegler and Kung–Schenck.

Contents

1. Introduction

One of the themes in the study of hyperplane arrangements is the relationship between their geometric properties and the combinatorics of their underlying matroids. In this article, we revisit the logarithmic differentials on a hyperplane arrangement and their relationship with the critical set or maximum likelihood variety of the arrangement.

We consider these objects from a unifying point of view that brings to light some new algebra and geometry of matroid realizations. To be more precise, let M be a matroid on a ground set E, and $V = \Bbbk^E$ the $\Bbbk$-vector space on E. A linear realization of M over a field $\Bbbk$ is a linear subspace $W \subseteq V$ for which the rank of a subset $S \subseteq E$ in M equals the rank of the projection of W onto the coordinate subspace $\Bbbk^S$. Let $W^\perp$ denote the $\Bbbk$-dual of V/W, a subspace of V^*. Then $W^\perp$ is a linear realization of the dual matroid, $\mathsf{M}^\perp$. The graph of the evaluation pairing

2020 *Mathematics Subject Classification.* Primary 05B35; Secondary 52C35.
Key words and phrases. Hyperplane arrangement.
The first author was partially supported by an NSERC Discovery Grant (Canada).

is, by definition, the map $V^* \times V \to V \times V$ given by $(\varphi, v) \mapsto (v, \varphi(v))$. Let $\mathcal{X}(W)$ denote the image of its restriction to $W \times W^\perp$.

In slightly different language, it was noted in [**Den14**, §5.3], [**HS14**] that $\mathcal{X}(W)$ is the affine critical set variety associated with W. This is a construction that arose both in mathematical physics applications of hyperplane arrangements [**Var95, OT95b**], as well as in maximal likelihood estimation in algebraic statistics [**CHKS06**]. We recall that if M has no loops, a linear realization W of M determines a hyperplane arrangement $\mathcal{A} = \{H_1, \ldots, H_n\}$ in W, where H_i is the intersection of the ith coordinate hyperplane of V with W. Here, M is called the matroid of the arrangement $\mathcal{A}$, and all hyperplane arrangements arise in this way. Maximum likelihood estimation motivates the problem of determining the zero locus of the logarithmic form $\omega_\lambda := d\log\left(\prod_{i=1}^n f_i^{\lambda_i}\right)$, where $\lambda \in \mathbb{Z}^n$ is a lattice vector and each f_i is a linear function on a finite-dimensional vector space W. For any f_i's and generic λ, Orlik and Terao [**OT95b**] proved a conjecture of Varchenko [**Var95**], that the (projective) zero locus consists of $\beta(\mathsf{M})$-many isolated points, where $\beta(\mathsf{M})$ is Crapo's β invariant.

A connection between $\mathcal{X}(W)$ and logarithmic forms appeared first in [**CDFV11**]. There, it is shown that $\mathcal{X}(W)$ is cut out set-theoretically by an ideal $I_{\log}$ generated by applying logarithmic differentials to a certain 1-form. The ideal is a complete intersection if and only if $\mathcal{A}$ is a free arrangement: that is, the module of logarithmic derivations $\mathrm{Der}(\mathcal{A})$ is free. It is shown in [**CDFV11**] that the ideal $I_{\log}$ is arithmetically Cohen–Macaulay under the weaker homological hypothesis that $\mathcal{A}$ is a tame arrangement. In general, $I_{\log}$ need not be radical. The relationship between the matroid M, the variety $\mathcal{X}(W)$ and homological properties of $\mathrm{Der}(\mathcal{A})$ is somewhat delicate, and a complete understanding would also settle Terao's long-standing Freeness Conjecture.

Another motivation comes from recent work of Ardila, Denham and Huh [**ADH23**] in which a key ingredient, the conormal fan of a matroid, is a tropical analogue of the variety $\mathcal{X}(W)$. A main result there [**ADH23**, Thm. 1.2] is a substantial generalization of Orlik and Terao's proof of Varchenko's conjecture, mentioned above. With this in mind, we feel that further investigation could also lead to additional combinatorial applications down the road.

1.1. Organization and main results. In §2, we review notation for arrangements, matroids and critical sets. For each matroid realization W, we consider the ideal $\mathfrak{a}$ generated by the pairwise products of linear forms $(f_1, \ldots, f_n)$ defining W and linear forms $(g_1, \ldots, g_n)$ defining $W^\perp$. We call this the *ideal of pairs*. We observe in §3 that it is closely related to the defining ideal $I_{\mathcal{X}}$ of the critical set variety $\mathcal{X}(W)$, though it has the advantage of being symmetric under matroid duality. In §2.4, we recall the notion of logarithmic forms on hyperplane arrangements, and we relate graded submodules of the ideals $I_{\log} \subseteq I_{\mathcal{X}}$ with logarithmic derivations (Proposition 2.4.2), strengthening a result from [**CDFV11**].

Our main results appear in Section §3. In §3.1, we show that logarithmic derivations for $\mathcal{A}$ and the dual arrangement $\mathcal{A}^\perp$ appear as certain syzygies of the ideal $\mathfrak{a}$. The affine variety $V(\mathfrak{a})$ is a subspace arrangement (by a result of Derksen and Sidman [**DS04**, Prop. 4.5]). In §3.2, we give a combinatorial description of those subspaces: it turns out that they are products of linear spaces from W and $W^\perp$, respectively, corresponding to a cyclic flat from $\mathcal{A}$ and the complementary cyclic flat from $\mathcal{A}^\perp$ (Theorem 3.2.4).

In general, though, the ideal $\mathfrak{a}$ is not reduced, and while we understand the minimal primes of $\mathfrak{a}$, the embedded primes remain somewhat mysterious. We give some examples, and in §3.3, we characterize what happens for uniform matroids. In §3.4, we use the geometry of $V(\mathfrak{a})$ to give an easy condition for an isomorphism type of the module of logarithmic derivations to change for different realizations of the same matroid.

As another application, in §3.5 we recall necessary combinatorial conditions for $\mathrm{Der}(\mathcal{A})$ to be free due to Kung and Schenck [**KS06**] and Ziegler [**Zie89**], and we show that they follow in a straightforward way from our description of $V(\mathfrak{a})$. Finally, we return to the question of when the inclusion $I_{\log} \subseteq I_X$ is an equality, and we show (Proposition 3.6.1) that this is the case when $\mathfrak{a}$ is an ideal of linear type.

2. Preliminaries

Let M be a matroid of rank r on the ground set $[n] := \{1, 2, \ldots, n\}$, and let $f \colon W \hookrightarrow \Bbbk^{[n]}$ be the inclusion of an r-dimensional linear subspace over a field $\Bbbk$. By definition, W is a *($\Bbbk$-)linear realization* of M provided the composite with each coordinate projection,

$$W \to \Bbbk^{[n]} \twoheadrightarrow \Bbbk^B,$$

is an isomorphism if and only if $B \subseteq [n]$ is a basis of M.

Let g denote the natural embedding of

$$W^\perp := (\Bbbk^{[n]}/W)^*$$

into the dual space $(\Bbbk^{[n]})^*$. $W^\perp$ is a linear realization of the dual matroid $\mathsf{M}^\perp$. We refer to [**Oxl11**] for terminology and background on matroid theory.

The remainder of this section proceeds as follows: In §2.1, we fix some additional notation that will be used throughout this article. In §2.2, we recall the definition of the hyperplane arrangement $\mathcal{A}$ associated to the realization W. In Proposition 2.3.1 of §2.3, we show how to represent the critical set variety of an arrangement as the spectrum of a subalgebra of $R \otimes_\Bbbk R^\perp$. This representation will be used in §3 as motivation to define the main tool of this paper: the ideal of pairs $\mathfrak{a}$. In §2.4, we relate the module of derivations $\mathrm{Der}(\mathcal{A})$ with certain ideals $I_{\log}$ and I_X—this relationship will be reformulated in terms of the ideal of pairs at the start of §3.1. We end this section with §2.5, which contains some technical commutative algebra results which will be needed later on.

2.1. Notation for some rings.

We make the following definitions:

- $x_1, \ldots, x_n$ – the coordinate functions on $\Bbbk^{[n]}$
- $y_1, \ldots, y_n$ – the dual coordinate functions on $(\Bbbk^{[n]})^*$
- $S := \Bbbk[x_1, \ldots, x_n]$ – the coordinate ring of $\Bbbk^{[n]}$
- $\mathbb{T}^n$ – the torus in $\Bbbk^{[n]}$ given by the equation $x_1 \cdots x_n \neq 0$
- f_i – the restriction of x_i to W for each $1 \le i \le n$
- g_i – the restriction of y_i to $W^\perp$ for each $1 \le i \le n$
- $R := \Bbbk[W] = \Bbbk[f_1, \ldots, f_n]$ and $R^\perp := \Bbbk[W^\perp] = \Bbbk[g_1, \ldots, g_n]$

Since $\Bbbk^{[n]} \cong W \oplus W^\perp$, we have $S = R \otimes_\Bbbk R^\perp$. We will also make use of a ring of parameters

$$A := \Bbbk[a_1, \ldots, a_n],$$

writing $R[a]$ and $S[a]$ interchangably for $R \otimes_{\Bbbk} A$ and $S \otimes_{\Bbbk} A$, respectively. The standard gradings on R, $R^{\perp}$, and A give multigradings: in order to distinguish the bigrading on S from that of $R[a]$, we will write the A-degree last and separate it with a semicolon.

2.2. Hyperplane arrangements. If M has no loops, then each $f_i \neq 0$, and we obtain a hyperplane arrangement $\mathcal{A}$ from W with hyperplanes

$$\{H_i := \ker f_i : 1 \leq i \leq n\}\,.$$

Similarly, if M has no coloops, then each $g_i \neq 0$, and we obtain a dual hyperplane arrangement $\mathcal{A}^{\perp}$ from $W^{\perp}$ in the same way. Given a hyperplane arrangement $\mathcal{A}$, we let

$$\begin{aligned} U(\mathcal{A}) &:= W - \bigcup_{i \in [n]} H_i \\ &= W \cap \mathbb{T}^n \end{aligned}$$

denote its complement, and let $\mathbb{P}U(\mathcal{A}) \subseteq \mathbb{P}W$ denote its quotient by the diagonal action of $\Bbbk^{\times}$.

EXAMPLE 2.2.1. Let W be the $\mathbb{C}$-span of the rows of the matrix

$$A := \begin{pmatrix} 1 & 0 & 0 & 1 & 1 \\ 0 & 1 & 0 & 1 & 0 \\ 0 & 0 & 1 & 0 & 1 \end{pmatrix},$$

a subspace of $\mathbb{C}^5$. Then $f = (x_1, x_2, x_3, x_1 + x_2, x_1 + x_3)$. The subspace W realizes a matroid M whose bases are, equivalently, the subsets of $[5]$ that index linearly independent columns of A, or linearly independent forms amongst the coordinates of f. The hyperplane arrangement $\mathcal{A} = \{H_1, \ldots, H_5\}$ is obtained by pulling back the coordinate hyperplanes from $\mathbb{C}^5$ to W.

The dual $\mathsf{M}^{\perp}$ is realized by the complement $W^{\perp}$. Choosing a basis, we may express $W^{\perp}$ as the $\mathbb{C}$-span of the rows of

$$A^{\perp} := \begin{pmatrix} -1 & -1 & 0 & 1 & 0 \\ -1 & 0 & -1 & 0 & 1 \end{pmatrix},$$

so that $g = (-y_1 - y_2, -y_1, -y_2, y_1, y_2)$. The corresponding rank-2 hyperplane arrangement has three distinct hyperplanes that appear with multiplicities 1, 2, and 2, respectively. $\diamondsuit$

2.3. The critical set variety. From various points of view it is of interest to consider rational functions with poles and zeroes on hyperplanes. Here, we take $\Bbbk = \mathbb{C}$.

Given a lattice vector $\lambda \in \mathbb{Z}^{[n]}$ and a realization $f \colon W \to \mathbb{C}^{[n]}$ of a matroid without loops, let

$$\Phi_{\lambda} := \prod_{i \in [n]} f_i^{\lambda_i},$$

regarded as a function $\Phi_{\lambda} \colon U(\mathcal{A}) \to \mathbb{C}^{\times}$. If $\sum_i \lambda_i = 0$, then Φ_{λ} induces a well-defined function on $\mathbb{P}U(\mathcal{A})$ as well.

By definition, the (affine) critical set variety $\mathfrak{X}(W) \subseteq W \times \mathbb{C}^{[n]} \subseteq \mathbb{C}^{[n]} \times \mathbb{C}^{[n]}$ is the Zariski closure of the pairs (p, λ) for which p is a critical point of the function Φ_{λ}. We refer to [**CDFV11**, **DGS12**] and [**Huh13**] for more details. The critical set variety is also an instance of a maximal likelihood variety—see, for example, [**HS14**].

If one restricts p to $U(\mathcal{A})$, then p is a critical point of Φ_λ if and only if the 1-form

$$\omega_\lambda := d\log\Phi_\lambda = \sum_{i=1}^{n}\lambda_i df_i/f_i$$

vanishes at p. A calculation shows that this is the case precisely when $(\lambda_1/f_1(p),\ldots,\lambda_n/f_n(p))$ is in $W^\perp$, which is to say that there exists a q such that for all i, $\lambda_i = f_i(p)g_i(q)$. Then $\mathfrak{X}(W)\cap(\mathbb{T}^n\times\mathbb{T}^n)$ is parameterized by

$$(f_1(p), f_2(p),\ldots,f_n(p), f_1(p)g_1(q),\ldots,f_n(p)g_n(q)) \tag{1}$$

for $(p,q)\in W\times W^\perp$.

PROPOSITION 2.3.1. *For any realization W of a matroid without loops, we have*

$$\begin{aligned}\mathfrak{X}(W) &= \operatorname{Spec}\mathbb{C}[f_1,\ldots,f_n,f_1g_1,\ldots,f_ng_n]\\ &= \operatorname{Spec}R[f_1g_1,\ldots,f_ng_n].\end{aligned}$$

PROOF. Let $\mathfrak{X}'$ denote the spectrum of the domain $R[f_1g_1,\ldots,f_ng_n]$, viewed as a closed subscheme of $\mathbb{C}^{[n]}\times\mathbb{C}^{[n]}$ via the $\mathbb{C}$-algebra map $\mathbb{C}[s_1,\ldots,s_n,t_1,\ldots,t_n]\to R[f_1g_1,\ldots,f_ng_n]$ induced by $s_i\mapsto f_i$ and $t_i\mapsto f_ig_i$ $(i=1,\ldots,n)$. The intersections of $\mathfrak{X}'$ and $\mathfrak{X}(W)$ with $\mathbb{T}^n\times\mathbb{T}^n$ agree. By construction, the torus $\mathbb{T}^n\times\mathbb{T}^n$ is dense in $\mathfrak{X}(W)$: taking closures, we see $\mathfrak{X}'$ contains $\mathfrak{X}(W)$. Clearly $\mathfrak{X}'$ is irreducible. By inverting each f_i we see $\dim\mathfrak{X}' = r+(n-r) = n$. On the other hand, $\mathfrak{X}(W)$ is also irreducible by [**CDFV11**, Cor. 2.10]) and has the same dimension, so they are equal. □

If W realizes a matroid with loops, we can take the equality above as the definition of $\mathfrak{X}(W)$. Clearly, deleting the loops leaves the algebra unchanged. We can also define $\mathfrak{X}(W)$ over an arbitrary field $\Bbbk$ in this way.

The defining ideal of $\mathfrak{X}(W)$ will be important in what follows.

DEFINITION 2.3.2. The subalgebra $R[f_1g_1,\ldots,f_ng_n]$ above is the image of a ring homomorphism $R[a]\to S$ given by sending $a_i\mapsto f_ig_i$ for $1\le i\le n$. We let $I_{\mathfrak{X}}$ denote the kernel, a prime ideal of codimension n in $R[a]$. We note that $I_{\mathfrak{X}}$ inherits the standard bigrading of $R[a]$.

2.4. Critical sets and logarithmic forms. Let $\operatorname{Der}(\mathcal{A})$ denote the R-module of logarithmic derivations on $\mathcal{A}$, and $\Omega^1(\mathcal{A}) = \operatorname{Der}(\mathcal{A})^\vee$ the dual module of logarithmic 1-forms — see, e.g., [**OT92**] for details. The logarithmic forms appear naturally in relation to the critical set variety. We will grade derivations so that $\deg(\partial/\partial x_i) = -1$ for each i.

THEOREM 2.4.1 ([**CDFV11**]). *The variety $\mathfrak{X}(W)$ is the zero set of the ideal of $R[a]$ given by*

$$I_{\log} := (\langle\theta,\omega_a\rangle : \theta\in\operatorname{Der}(\mathcal{A}))\,,$$

where $\omega_a := \sum_i a_i df_i/f_i$.[1]

[1]Equivalently, $I_{\log}$ is generated by applying logarithmic derivations to the formal expression $\log\left(\prod_{i=1}^n f_i^{a_i}\right)$.

Recall that an arrangement $\mathcal{A}$ is said to be *free* if $\mathrm{Der}(\mathcal{A})$ is a free R-module. In [**CDFV11**], it is shown that $\mathcal{A}$ is free if and only if $I_{\log}$ is a complete intersection. A weaker but quite useful property goes back to [**OT95a**]: an arrangement $\mathcal{A}$ is *tame* if $\operatorname{pdim} \Omega^p(\mathcal{A}) \leq p$ for $1 \leq p \leq r$, where $\Omega^p(\mathcal{A}) \cong \left(\bigwedge^p \Omega^1(\mathcal{A})\right)^{\vee\vee}$ is the module of logarithmic p-forms (see [**DS12**].)

The Theorem shows $I_{\log} \subseteq I_{\mathcal{X}}$, but equality fails in general — see Example 2.4.3 below. However, if $\mathcal{A}$ is tame, then $I_{\log} = I_{\mathcal{X}}$ ([**CDFV11**]).

PROPOSITION 2.4.2. *As graded R-modules,*

$$(I_{\log})_{(\cdot;1)} = (I_{\mathcal{X}})_{(\cdot;1)} \cong \mathrm{Der}(\mathcal{A})(-1).$$

PROOF. Noting that $I_{\log} \subseteq I_{\mathcal{X}}$, we will construct maps $(I_{\mathcal{X}})_{(p;1)} \to \mathrm{Der}(\mathcal{A})_{p-1} \to (I_{\log})_{(p;1)}$ which compose to the inclusion $I_{\mathcal{X}} \subseteq I_{\log}$, for each integer p. Any $u \in (I_{\mathcal{X}})_{(p;1)}$ can be written

$$u = \sum_{i=1}^{n} c_i a_i \quad \text{where} \quad \sum_{i=1}^{n} c_i f_i g_i = 0, \tag{2}$$

for some homogeneous elements $c_i \in R$. By reordering the ground set of M if necessary, we assume that $[r]$ is independent in M, so that we may choose bases for W and $W^{\perp}$, respectively, with $f_i = x_i$ for $1 \leq i \leq r$ and $g_i = x_i$ for $r+1 \leq i \leq n$.

Let $\theta = \sum_{i=1}^{r} c_i x_i \partial/\partial x_i$. We claim $\theta(f_j) = c_j f_j$ for all j, so $\theta \in \mathrm{Der}(\mathcal{A})$. If so,

$$\langle \theta, \omega_a \rangle = \sum_{i=1}^{n} a_i \theta(f_i)/f_i = u,$$

which implies $u \in I_{\log}$.

For $1 \leq j \leq r$ the claim is obvious. For $r+1 \leq j \leq n$, we compute as follows. By orthogonality, $\partial f_j/\partial x_i = -\partial g_i/\partial x_j$ for all $1 \leq i \leq r$ and $r+1 \leq j \leq n$. Differentiating (2) by x_j shows, for each j,

$$\begin{aligned} c_j f_j &= -\sum_{i=1}^{r} c_i f_i \frac{\partial g_i}{\partial x_j} \\ &= \sum_{i=1}^{r} c_i f_i \frac{\partial f_j}{\partial x_i} \\ &= \theta(f_j), \end{aligned}$$

as required. □

EXAMPLE 2.4.3. Consider the arrangement $\mathcal{A}$ of 9 hyperplanes in $\mathbb{C}^4$ defined by the columns of the matrix

$$\begin{pmatrix} 1 & 0 & 0 & 1 & 0 & 0 & 1 & 1 & 0 \\ 0 & 1 & 0 & 0 & 1 & 0 & 1 & 0 & 1 \\ 0 & 0 & 1 & 0 & 0 & 1 & 0 & 1 & 1 \\ 0 & 0 & 0 & 1 & 1 & 1 & 1 & 1 & 1 \end{pmatrix}.$$

This arrangement has $\operatorname{pdim} \Omega^1(\mathcal{A}) = 2$, so it is not tame. A calculation with Macaulay2 [**GS**] shows that $I_{\log}$ has an embedded prime at the origin.

Clearly $I_{\log}$ is always generated by elements of bidegree $(p;1)$, for integers $p \geq 0$. For this arrangement, however, computation also shows that $I_{\mathcal{X}}$ has a generator of degree $(2;2)$. ◊

2.5. Associated primes and slicing. We end this section with some technical results which we will use to prove Corollary 3.2.12.

LEMMA 2.5.1. *Let S be a bigraded Noetherian ring, $R := S_{(\cdot,0)}$. Let $k \in \mathbb{Z}$, and let M be a graded S-module. Let $\mathfrak{p}$ be a homogeneous prime ideal in S. If N is a $\mathfrak{p}$-primary graded submodule of M and $N_{(\cdot,k)} \neq M_{(\cdot,k)}$, then $N_{(\cdot,k)}$ is a $\mathfrak{p}_{(\cdot,0)}$-primary submodule of $M_{(\cdot,k)}$.*

PROOF. Let $x \in M_{(\cdot,k)}$ with $x \notin N_{(\cdot,k)}$, and let $a \in R$ be homogeneous. Suppose $ax \in N_{(\cdot,k)}$. Then $ax \in N$, so since N is $\mathfrak{p}$-primary and $x \notin N$, $a \in \mathfrak{p}$. Thus, $a \in \mathfrak{p}_{(\cdot,0)}$. Hence, $N_{(\cdot,k)}$ is $\mathfrak{p}_{(\cdot,0)}$-primary. □

EXAMPLE 2.5.2. The condition $N_{(\cdot,k)} \neq M_{(\cdot,k)}$ in Lemma 2.5.1 is necessary: Let $S = \mathbb{C}[x,y]$ be standard bigraded, $M = S$, and $N = \mathfrak{p} = (x,y)$. Then $M_{(\cdot,1)} = N_{(\cdot,1)} = R$, so in particular $M_{(\cdot,1)}/N_{(\cdot,1)}$ has no associated primes. ◊

COROLLARY 2.5.3. *Let S be a bigraded Noetherian ring, $R := S_{(\cdot,0)}$. Let $k \in \mathbb{Z}$, let M be a graded S-module, let $\mathfrak{p}_1, \dots, \mathfrak{p}_s$ be the associated primes of M, and let $N_1 \cap \cdots \cap N_s$ be a minimal (homogeneous) primary decomposition of 0 in M, where N_i is $\mathfrak{p}_i$-primary.*

(a) *The associated primes of $M_{(\cdot,k)}$ are contained in $\{(\mathfrak{p}_i)_{(\cdot,0)} : (N_i)_{(\cdot,k)} \neq M_{(\cdot,k)}\}$, hence also in $\{\mathfrak{p}_{(\cdot,0)} : \mathfrak{p}$ an associated prime of $M\}$.*
(b) *The minimal primes of $M_{(\cdot,k)}$ are $\min\{(\mathfrak{p}_i)_{(\cdot,0)} : (N_i)_{(\cdot,k)} \neq M_{(\cdot,k)}\}$.*

PROOF. By Lemma 2.5.1, $\bigcap_i (N_i)_{(\cdot,k)}$, where i runs through those indices for which $(N_i)_{(\cdot,k)} \neq M_{(\cdot,k)}$, is a (not necessarily minimal) primary decomposition of 0 in $M_{(\cdot,k)}$. Now use standard facts relating the associated primes of M to primary decompositions of 0 in M to get (a).

We now prove (b). A standard fact about minimal primes says that if L is a finitely-generated R-module, $\mathfrak{q}_1, \dots, \mathfrak{q}_t$ are (not necessarily distinct) primes in R, and $K_1 \cap \cdots \cap K_t$ is a not necessarily minimal primary decomposition of 0 in L, then the minimal primes of L are exactly the minimal elements of $\{\mathfrak{q}_1, \dots, \mathfrak{q}_t\}$. Now use that $\bigcap_i (N_i)_{(\cdot,k)}$, where i runs through those indices for which $(N_i)_{(\cdot,k)} \neq M_{(\cdot,k)}$, is a primary decomposition of 0 in $M_{(\cdot,k)}$, and that the relevant $(N_i)_{(\cdot,k)}$ are $(\mathfrak{p}_i)_{(\cdot,0)}$-primary. □

3. The ideal of pairs

In this section, we introduce and study the ideal of pairs associated with a pair of matroid realizations $(W, W^{\perp})$. This ideal determines a subspace arrangement which we are able to understand set-theoretically but not scheme-theoretically. Syzygies of the ideal of pairs contain information about logarithmic derivations on both hyperplane arrangements $\mathcal{A}$ and $\mathcal{A}^{\perp}$.

DEFINITION 3.0.1. The *ideal of pairs* associated to a realization W is the ideal

$$\mathfrak{a} := \mathfrak{a}_{W,W^{\perp}} := (f_1 g_1, \dots, f_n g_n)$$

of S.

Clearly,

$$\mathfrak{a}_{(\cdot,1)} \cong \bigoplus_{i \geq 0} (R[a]/I_X)_{(i;1)} = (R[a]/I_X)_{(\cdot;1)}.$$

3.1. Syzygies of $\mathfrak{a}$. Relations amongst the products $\{f_i g_i\}$ are related to the discussion in Section §2.4, and Proposition 2.4.2 can be reformulated as follows. Let K denote the kernel of the homomorphism $S[a]_{(\cdot,\cdot;1)}(-1,-1) \to \mathfrak{a}$ sending a_i to $f_i g_i$.

THEOREM 3.1.1. *Let W be a subspace realizing a matroid M. Let Π be the partition of $[n]$ given by the connected components of M.*

- *$K_{(1,1)}$ is spanned by sums $\sum_{i\in C} a_i$, for each connected component C in Π.*
- *$K_{(\cdot,1)} \cong \mathrm{Der}(\mathcal{A})(-1)$ as graded R-modules, and*
- *$K_{(1,\cdot)} \cong \mathrm{Der}(\mathcal{A}^\perp)(-1)$ as graded $R^\perp$-modules.*

PROOF. As R-modules, we have $(I_X)_{(\cdot;1)} \cong K_{(\cdot,1)}$ by construction. So the second claim follows by Proposition 2.4.2, and the third claim by exchanging the roles of R and $R^\perp$.

To establish the first claim, we note that the only degree-zero derivations on a connected component C are multiples of the Euler derivation, $\sum_{i\in C} x_i\partial/\partial x_i$. Now $\mathrm{Der}(\mathcal{A})$ is a direct sum of logarithmic derivations on the connected components (see, e.g., [**OT92**]), and the isomorphism $K_{(1,1)} \cong \mathrm{Der}(\mathcal{A})_0$ sends a_i to $x_i\partial/\partial x_i$. □

3.1.1. *Minimal syzygies.* The claim about $K_{(\cdot,1)}$ in Theorem 3.1.1 can be reformulated as saying that we have exact sequences

$$0 \longrightarrow \mathrm{Der}(\mathcal{A})(-1) \longrightarrow F \longrightarrow \mathfrak{a}_{(\cdot,1)} \longrightarrow 0, \tag{3}$$

and

$$0 \longrightarrow \mathrm{Der}(\mathcal{A})(-1) \longrightarrow F \longrightarrow E \longrightarrow (S/\mathfrak{a})_{(\cdot,1)} \longrightarrow 0 \tag{4}$$

where $E = S_{(\cdot,1)} \cong R \otimes_{\Bbbk} (W^\perp)^*$ is a free R-module of rank $n-r$, and $F = S[a]_{(\cdot,0;1)}(-1) = R[a]_{(\cdot,1)}(-1)$ is a free R-module of rank n.

We can refine these exact sequences slightly using Theorem 3.1.1. Let $\theta_1,\ldots,\theta_\kappa$ be the Euler derivations from the proof of Theorem 3.1.1. Define

$$\overline{\mathrm{Der}}(\mathcal{A}) := \frac{\mathrm{Der}(\mathcal{A})}{\sum_{i=1}^{\kappa} R\theta_i}.$$

REMARK 3.1.2. Assume for simplicity that M has no loops. When the characteristic of $\Bbbk$ does not divide the cardinality of any connected component of M, then $\overline{\mathrm{Der}}(\mathcal{A})$ is canonically isomorphic to the submodule

$$\mathrm{Der}_0(\mathcal{A}) := \{\theta \in \mathrm{Der}(\mathcal{A}) : \theta(f_1\cdots f_n) = 0\}$$

of $\mathrm{Der}(\mathcal{A})$: If n_i is the cardinality of the ith connected component of M, and Q_i is the defining polynomial of the ith connected component of the arrangement, then

$$\theta \mapsto \theta - \sum_{i=1}^{\kappa} \frac{1}{n_i}\frac{\theta(Q_i)}{Q_i}\theta_i$$

induces a well-defined map from $\overline{\mathrm{Der}}(\mathcal{A})$ isomorphically to $\mathrm{Der}_0(\mathcal{A})$.

The assumption on the characteristic is necessary: For instance, if $f = (x, y, x+y)$ and the characteristic of $\Bbbk$ is 3, then $\mathrm{Der}_0(\mathcal{A}) = \mathrm{Der}(\mathcal{A})$, which is certainly not isomorphic to $\overline{\mathrm{Der}}(\mathcal{A})$ (the two have different ranks). ◊

Recall that a graded R-module N is a *minimal graded kth syzygy module* of the graded R-module M if there exists an exact sequence

$$\xi\colon N \xrightarrow{\partial_{k+1}} P_k \xrightarrow{\partial_k} \cdots \xrightarrow{\partial_1} P_0$$

such that

- each P_i is graded free;
- $\operatorname{coker}(\partial_1) \cong M$; and
- the differentials of $\Bbbk \otimes_R \xi$ are all zero, or equivalently, $\operatorname{im}(\partial_{i+1}) \subseteq R_+ P_i$ for each $0 \leq i \leq k$, where R_+ is the homogeneous maximal ideal.

LEMMA 3.1.3.

(a) *The graded R-module $F/RK_{(1,1)}$ is graded free, where F is the graded free R-module defined above, and K is the module defined at the start of* §3.1.

(b) *There are exact sequences*

$$0 \longrightarrow \overline{\operatorname{Der}}(\mathcal{A})(-1) \longrightarrow F/RK_{(1,1)} \longrightarrow \mathfrak{a}_{(\cdot,1)} \longrightarrow 0, \tag{5}$$

and

$$0 \longrightarrow \overline{\operatorname{Der}}(\mathcal{A})(-1) \longrightarrow F/RK_{(1,1)} \longrightarrow E \longrightarrow (S/\mathfrak{a})_{(\cdot,1)} \longrightarrow 0 \tag{6}$$

where E and F are the graded free R-modules defined above.

(c) $\overline{\operatorname{Der}}(\mathcal{A})(-1)$ *is a minimal graded first and second syzygy module of $\mathfrak{a}_{(\cdot,1)}$ and $(S/\mathfrak{a})_{(\cdot,1)}$, respectively.*

PROOF. (a) By Theorem 3.1.1, the vector space $K_{(1,1)}$ is spanned by the images of the Euler derivations $\theta_1, \ldots, \theta_\kappa$ corresponding to the connected components of M. These images all live in the vector space F_1 spanned by $a_1, \ldots, a_n$. Hence, $F/RK_{(1,1)} \cong R(-1) \otimes_{\Bbbk} (F_1/K_{(1,1)})$ is a graded free R-module.

(b) By Theorem 3.1.1, the vector space $K_{(1,1)}$ is in the kernel of the map $F \to \mathfrak{a}_{(\cdot,1)}$, and it is spanned by the images of the Euler derivations $\theta_1, \ldots, \theta_\kappa$ corresponding to the connected components of M. Now use the definition of $\overline{\operatorname{Der}}(\mathcal{A})$.

(c) Every degree zero element of $\operatorname{Der}(\mathcal{A})$ lives in the R-span of the Euler derivations. Therefore, the image of $\overline{\operatorname{Der}}(\mathcal{A})(-1)$ in $F/RK_{(1,1)}$ lives in degrees at least 1, i.e. in $R_+ \cdot (F/RK_{(1,1)})$, where R_+ is the homogeneous maximal ideal. This shows that $\overline{\operatorname{Der}}(\mathcal{A})$ is a minimal graded first syzygy module of $\mathfrak{a}_{(\cdot,1)}$. The claim about $(S/\mathfrak{a})_{(\cdot,1)}$ follows once we observe that, by definition, the image $\mathfrak{a}_{(\cdot,1)}$ of $F/RK_{(1,1)} \to E$ is contained in R_+E. □

REMARK 3.1.4. If A is free, then the proof of Lemma 3.1.3(c) shows that (5) and (6) are minimal graded free resolutions of $\mathfrak{a}_{(\cdot,1)}$ and $(S/\mathfrak{a})_{(\cdot,1)}$, respectively. ◊

3.1.2. *Tor of the module of derivations.* Recall that W and $W^\perp$ are r- and $(n-r)$-dimensional, respectively.

COROLLARY 3.1.5. *For all $p \geq 1$ and $i \geq 1$, we have*

$$\operatorname{Tor}^S_{p+1}(\mathfrak{a}, \Bbbk)_{(i,1)} \cong \operatorname{Tor}^R_p(\operatorname{Der}(\mathcal{A}), \Bbbk)_{i-1} \quad \textit{and}$$
$$\operatorname{Tor}^S_{p+1}(\mathfrak{a}, \Bbbk)_{(1,i)} \cong \operatorname{Tor}^{R^\perp}_p(\operatorname{Der}(\mathcal{A}^\perp), \Bbbk)_{i-1}.$$

Additionally, for all $i \geq 1$, we have

$$\dim_{\Bbbk} \operatorname{Tor}_1^S(\mathfrak{a}, \Bbbk)_{(i,1)} = \dim_{\Bbbk}(\operatorname{Der}(\mathcal{A}) \otimes_R \Bbbk)_{i-1} - \binom{i+r-2}{r-1}\kappa,$$

$$\dim_{\Bbbk} \operatorname{Tor}_1^S(\mathfrak{a}, \Bbbk)_{(1,i)} = \dim_{\Bbbk}(\operatorname{Der}(\mathcal{A}^{\perp}) \otimes_R \Bbbk)_{i-1} - \binom{i+n-r-2}{n-r-1}\kappa,$$

where κ is the number of components of M.

PROOF. If F is a graded free S-module, then $F_{(\cdot,1)}$ is a graded free R-module. Moreover, $(-)_{(\cdot,1)}$ is an exact functor. Thus, $(-)_{(\cdot,1)}$ takes graded free resolutions to graded free resolutions. Thus, the statement for $p \geq 1$ follows from (3). The statement for $p = 0$ follows from (5), the definition of $\overline{\operatorname{Der}}(\mathcal{A})$, and the formula for the Hilbert function of the polynomial rings R and $R^{\perp}$. □

Thus the ideal of pairs "sees" in particular whether or not $\mathcal{A}$ or its dual are free arrangements.

EXAMPLE 3.1.6. Let $\mathcal{A}$ be the braid arrangement A_3. This is also the rank-3 graphic arrangment defined by the complete graph K_4. Here $n = 6$, the arrangement $\mathcal{A}$ is free, and $\mathcal{A}^{\perp}$ is isomorphic to $\mathcal{A}$. The ideal $\mathfrak{a}$ has projective dimension 3, and its bigraded Betti numbers are given by:

$$\begin{array}{|lll} t & 5t^2 & t^3 \\ t & 7t & 5t^2 \\ 5 & t & t \\ \hline \end{array}\,, \tag{7}$$

where the (i, j) entry is the polynomial

$$\sum_{p \geq 0} \dim_{\Bbbk} \left(\operatorname{Tor}_p^S(\mathfrak{a}, \Bbbk)_{(i,j)} \right) t^p.$$

The root system A_3 has coexponents $\{1, 2, 3\}$, so $\operatorname{Der}(\mathcal{A}) \cong R \oplus R(-1) \oplus R(-2)$, and the two positive-degree generators of $\operatorname{Der}(\mathcal{A})$ are reflected in the bottom row of (7). ◊

3.2. Minimal primes over $\mathfrak{a}$. The ideal $\mathfrak{a}$ defines a variety $V(\mathfrak{a})$ in $\mathbb{A}^n$ whose components are all linear subspaces, by a result of Derksen and Sidman [**DS04**, Prop. 4.5]. We describe these linear subspaces in Theorem 3.2.4. We use this description in Corollary 3.2.9 to produce a combinatorial lower bound for $\operatorname{pdim}_S(S/\mathfrak{a})$, and then again in Corollary 3.2.12 to describe the minimal associated primes of $\mathfrak{a}_{(\cdot,1)}$.

Let $L(\mathsf{M})$ denote the lattice of flats of M (ordered by inclusion). It is well known that $L(\mathsf{M})$ is isomorphic to the lattice of subspaces cut out by the hyperplane arrangement $\mathcal{A}$ (ordered by reverse inclusion).

A flat $F \in L(\mathsf{M})$ is *cyclic* if F is a union of circuits or, equivalently, if its complement $F^{\mathsf{C}} \in L(\mathsf{M}^{\perp})$. Let $\mathcal{Z}(\mathsf{M}) \subseteq L(\mathsf{M})$ denote the subposet of cyclic flats. Here is a basic fact about cyclic flats. (See, for example, [**Oxl11**, Ex. 8.2.13], the closely related [**FHG**$^+$**21**, Cor. 6].)

PROPOSITION/DEFINITION 3.2.1. *For a flat* $F \in L(\mathsf{M})$, *let*

$$Z(F) := \bigcup_{C \subseteq F} C,$$

where C *ranges over all circuits of* M. *Then* $Z(F)$ *is a cyclic flat, and* $\mathrm{cl}_{\mathsf{M}^\perp}(F^{\mathsf{C}}) = Z(F)^{\mathsf{C}}$.

DEFINITION 3.2.2. We say a pair $(F, G) \in L(\mathsf{M}) \times L(\mathsf{M}^\perp)$ is a *biflat* if $F \cup G = [n]$.[2] Let $L(\mathsf{M}, \mathsf{M}^\perp)$ denote the subposet of $L(\mathsf{M}) \times L(\mathsf{M}^\perp)$ consisting of just the biflats.

For any subsets $I, J \subseteq [n]$, we let

$$\mathfrak{p}_{I,J} := (f_i \colon i \in I) + (g_j \colon j \in J),$$

a linear ideal of S, and let $L_{I,J} := V(\mathfrak{p}_{I,J})$.

REMARK 3.2.3. We have $\mathfrak{p}_{I,J} = \mathfrak{p}_{\mathrm{cl}_{\mathsf{M}}(I), \mathrm{cl}_{\mathsf{M}^\perp}(J)}$, since a linear functional f_j is in the span of $\{f_i \colon i \in I\}$ if and only if j is in the (matroid) span of I. Clearly

$$\operatorname{codim} \mathfrak{p}_{I,J} = \mathrm{rk}_{\mathsf{M}}(I) + \mathrm{rk}_{\mathsf{M}^\perp}(J). \qquad \diamondsuit$$

THEOREM 3.2.4. *Let* W *be a realization of a matroid* M *and* $\mathfrak{a}$ *its ideal of pairs.*

(a) *Every associated prime of* $S/\mathfrak{a}$ *is of the form* $\mathfrak{p}_{F,G}$ *for some biflat* (F, G).
(b) *The minimal primes of* $S/\mathfrak{a}$ *are* $\{\mathfrak{p}_{F,F^{\mathsf{C}}} \colon F \in \mathcal{Z}(\mathsf{M})\}$. *In particular,*

$$V(\mathfrak{a}) = \bigcup_{F \in \mathcal{Z}(\mathsf{M})} L_{F,F^{\mathsf{C}}}.$$

(c) *If* $F \in \mathcal{Z}(\mathsf{M})$, *the primary component of* $\mathfrak{a}$ *corresponding to* $\mathfrak{p}_{F,F^{\mathsf{C}}}$ *is* $\mathfrak{p}_{F,F^{\mathsf{C}}}$ *itself.*

PROOF. (a) Let P be an associated prime of $\mathfrak{a}$. By [**DS04**, Prop. 4.5], P is necessarily a linear ideal of the form $\mathfrak{p}_{F,G}$ for some subsets $F, G \subseteq [n]$. By the previous remark, we may assume $F \in L(\mathsf{M})$ and $G \in L(\mathsf{M}^\perp)$. That $F \cup G = [n]$ follows from the fact that P is prime and contains $f_i g_i$ for each i.

(b) Let P be a minimal prime of $S/\mathfrak{a}$. By (a), we have $P = \mathfrak{p}_{F,G}$ for some biflat (F, G). By Proposition 3.2.1, we have $Z(F) \subseteq F$ and $\mathrm{cl}_{\mathsf{M}^\perp}(F^{\mathsf{C}}) = Z(F)^{\mathsf{C}}$.

Since $F^{\mathsf{C}} \subseteq G$, taking closures in $\mathsf{M}^\perp$ shows $Z(F)^{\mathsf{C}} \subseteq G$ as well. By minimality, then, $F = Z(F)$ and $G = F^{\mathsf{C}}$.

(c) Localizing $\mathfrak{a}$ at $\mathfrak{p} = \mathfrak{p}_{F,F^{\mathsf{C}}}$ gives the ideal $\mathfrak{p}S_{\mathfrak{p}}$. Now we note that $\mathfrak{p}S_{\mathfrak{p}} \cap S = \mathfrak{p}$. $\square$

REMARK 3.2.5. If M has no loops or coloops, then $[n]$ is a cyclic flat. Therefore, the prime ideals $\mathfrak{p}_{[n],\emptyset} = \langle W^* \rangle$ and $\mathfrak{p}_{\emptyset,[n]} = \langle (W^\perp)^* \rangle$ are minimal primes of $S/\mathfrak{a}$ by Theorem 3.2.4(b). $\diamondsuit$

EXAMPLE 3.2.6. If M is a uniform matroid, then the associated primes of $S/\mathfrak{a}$ are $\mathfrak{p}_{[n],\emptyset}$, $\mathfrak{p}_{\emptyset,[n]}$, and $\mathfrak{p}_{[n],[n]}$. This will be proven in Corollary 3.3.3. $\diamondsuit$

EXAMPLE 3.2.7 (Example 2.4.3 continued). A computation using Macaulay2 [**GS**] shows that $S/\mathfrak{a}$ has no embedded primes. $\diamondsuit$

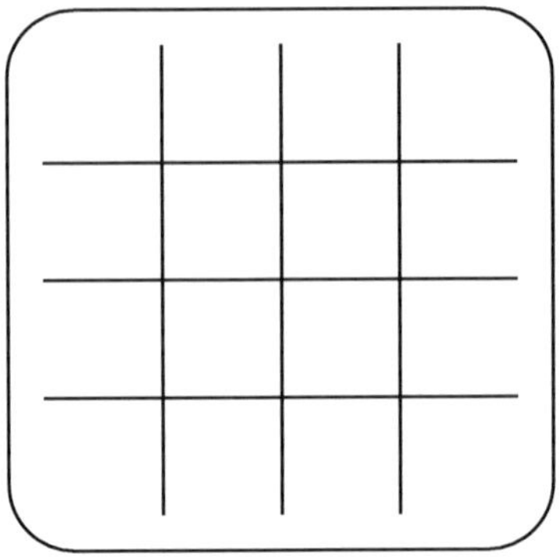

FIGURE 1. The projectivization of the arrangement from Example 3.2.8 with the hyperplane $f_1 = 0$ placed at infinity.

EXAMPLE 3.2.8. Consider the arrangement $\mathcal{A}$ of 7 hyperplanes in $\mathbb{C}^3$ defined by the columns of the matrix

$$\begin{pmatrix} 1 & 1 & 1 & 1 & 1 & 1 & 1 \\ 0 & 1 & 0 & 2 & 0 & 3 & 0 \\ 0 & 0 & 1 & 0 & 2 & 0 & 3 \end{pmatrix}.$$

This arrangement is pictured in Figure 1. A computation using Macaulay2 [**GS**] shows that the embedded primes correspond to the biflats $(1246, [7])$, $(1357, [7])$, and $([7], [7])$. ◊

COROLLARY 3.2.9. *For any arrangement $\mathcal{A}$, we have*

$$\operatorname{pdim}_S(S/\mathfrak{a}) \geq \max\{2\operatorname{rk}(F) - |F| + n - r \colon F \in \mathcal{Z}(\mathsf{M})\}$$

PROOF. Localization does not increase projective dimension, and the projective dimension of the residue field at $\mathfrak{p}_{F,F^{\mathsf{C}}}$ equals the codimension, which by [**Oxl11**, Prop. 2.1.9] is

$$\operatorname{rk}_{\mathsf{M}}(F) + \operatorname{rk}_{\mathsf{M}^\perp}(F^{\mathsf{C}}) = 2\operatorname{rk}_{\mathsf{M}}(F) - |F| + n - r. \qquad \square$$

We note that $\operatorname{pdim}_S(S/\mathfrak{a}) \geq n - r$: since we assume the underlying matroid has no loops, $F = \emptyset$ is a cyclic flat. If, additionally, M has no coloops, then $\operatorname{pdim}_S(S/\mathfrak{a}) \geq r$, since then $F = E \in \mathcal{Z}(\mathsf{M})$.

EXAMPLE 3.2.10 (Example 3.1.6 continued). The A_3 arrangement has four proper cyclic flats given by the four triangles in the complete graph K_4. They each have rank 2, and so do their complements, so $\operatorname{codim}\mathfrak{p}_{F,F^{\mathsf{C}}} = 2 + 2$ for proper cyclic flats F, and $\operatorname{codim}\mathfrak{p}_{F,F^{\mathsf{C}}} = 3 + 0$ for $F = \emptyset$ and $F = E$. So $\operatorname{pdim} S/\mathfrak{a} \geq 4$, and our previous computation shows the bound is sharp in this case. It is also interesting to note that according to a computation using Macaulay2 [**GS**], $S/\mathfrak{a}$ has no embedded primes. ◊

EXAMPLE 3.2.11 (Example 3.2.8 continued). The cyclic flats are $\emptyset$, 1246, 1357, and $[7]$, which have ranks 0, 2, 2, and 3, resp. Therefore,

$$\max\{2\operatorname{rk}(F) - |F| + n - r \colon F \in \mathcal{Z}(\mathsf{M})\} = \max\{4, 4, 4, 3\} = 4$$

[2]This disagrees slightly with the language of [**ADH23**], where in addition F and G must be nonempty and not both equal to $[n]$.

However, a computation using Macaulay2 [**GS**] shows that $\operatorname{pdim}_S(S/\mathfrak{a}) = 7$. $\diamondsuit$

For a subset $F \subseteq [n]$, let

$$P_F := (f_i : i \in F),$$

an ideal of R.

COROLLARY 3.2.12. *Let W be a matroid realization with no loops or coloops.*

(a) *The associated primes of $(S/\mathfrak{a})_{(\cdot,1)}$ are contained in $\{P_F : \emptyset \neq F \in L(\mathsf{M})\}$.*
(b) *The minimal primes of $(S/\mathfrak{a})_{(\cdot,1)}$ are $\{P_F : F \in \min(\mathcal{Z}(\mathsf{M}) \setminus \{\emptyset\})\}$.*

PROOF. This follows immediately from Theorem 3.2.4(a), (b), and (c) together with Corollary 2.5.3. $\square$

EXAMPLE 3.2.13. If M is a uniform matroid, then its only cyclic flats are $\emptyset$ and $[n]$. Therefore, by Corollary 3.2.12(b), the only associated prime of $(S/\mathfrak{a})_{(\cdot,1)}$ is $P_{[n]}$. This can also be proven using Lemma 3.3.1 below. $\diamondsuit$

EXAMPLE 3.2.14 (Example 3.2.7 continued). A computation using Macaulay2 [**GS**] shows that neither $(S/\mathfrak{a})_{(\cdot,1)}$ nor $(S/\mathfrak{a})_{(1,\cdot)}$ have any embedded primes. $\diamondsuit$

EXAMPLE 3.2.15 (Example 3.2.8 continued). A computation using Macaulay2 [**GS**] shows that, although $S/\mathfrak{a}$ has embedded primes, $(S/\mathfrak{a})_{(\cdot,1)}$ does not. On the other hand, $(S/\mathfrak{a})_{(1,\cdot)}$ has does have an embedded prime, and this embedded prime is at the origin. $\diamondsuit$

QUESTION 3.2.16.

(a) *What are the embedded primes of $(S/\mathfrak{a})_{(\cdot,1)}$?*
(b) *A priori, the flats F for which P_F is an embedded prime of $(S/\mathfrak{a})_{(\cdot,1)}$ depend on the choice of realization. Are they, in fact, determined by the matroid M?*

3.3. The uniform matroid. In Corollary 3.3.3 below, we prove the claim from Example 3.2.6 about the associated primes of $\mathfrak{a}$ when M is uniform. In order to do so, we will need the following lemma:

LEMMA 3.3.1. *If $\mathsf{M} = U_{r,n}$, then*

$$g_{i_1} \cdots g_{i_r} f_a \in \mathfrak{a}$$

for all $i_1, \ldots, i_r, a \in [n]$. In other words, $\mathfrak{p}^r_{\emptyset,[n]} W^* \subseteq \mathfrak{a}$.

PROOF. We proceed by induction on the number

$$\nu = \nu(i_1, \ldots, i_r) := \sum_{j \in [n]} \nu_j(i_1, \ldots, i_r),$$

where

$$\nu_j = \nu_j(i_1, \ldots, i_r) := \begin{cases} |\{k : i_k = j\}| - 1, & \text{if } j \in \{i_1, \ldots, i_r\}, \\ 0, & \text{otherwise.} \end{cases}$$

If $\nu = 0$, then all the i_j's are distinct, and therefore by uniformity f_a is in the $\Bbbk$-span of $f_{i_1}, \ldots, f_{i_r}$, i.e. $f_a = c_1 f_{i_1} + \cdots + c_r f_{i_r}$ for some $c_k \in \Bbbk$. So, $g_{i_1} \cdots g_{i_r} f_a \in \mathfrak{a}$.

Suppose $\nu > 0$. After re-indexing, we may assume that i_1 appears at least twice, i.e. $\nu_{i_1} > 0$. Let $j_1, \ldots, j_{n-r}$ be distinct elements of $[n] \setminus \{i_1, \ldots, i_r\}$. By uniformity, $\{j_1, \ldots, j_{n-r}\}$ is a basis of $\mathsf{M}^\perp$, and therefore g_{i_1} is in the $\Bbbk$-span of $g_{j_1}, \ldots, g_{j_{n-r}}$. So,

$$g_{i_1} \cdots g_{i_r} f_a \in \sum_{k=1}^{n-r} \Bbbk g_{j_k} g_{i_2} \cdots g_{i_r} f_a.$$

By construction, j_k appears exactly once in $j_k, i_2, \ldots, i_r$, and i_1 appears one fewer time than in $i_1, \ldots, i_r$. So, $\nu_{j_k}(j_k, i_2, \ldots, i_r) = 0$, $\nu_{i_1}(j_k, i_2, \ldots, i_r) = \nu_{i_1}(i_1, \ldots, i_r) - 1$, and no other ν_ℓ is affected. Thus,

$$\nu(j_k, i_2, \ldots, i_r) = \nu(i_1, i_2, \ldots, i_r) - 1.$$

Hence, by the induction hypothesis, $g_{j_k} g_{i_2} \cdots g_{i_r} f_a \in \mathfrak{a}$ for each k. Hence, $g_{i_1} \cdots g_{i_r} f_a \in \mathfrak{a}$. □

REMARK 3.3.2. In the proof of Lemma 3.3.1, we showed in particular that if $i_1, \ldots, i_r \in [n]$ are distinct, then $g_{i_1} \cdots g_{i_r} f_a \in \mathfrak{a}$ for every $a \in [n]$. The same argument proves a more general statement: If W is a realization of a matroid M and B is a basis of M, then for all $a \in [n]$,

$$\left(\prod_{i \in B} g_i\right) f_a \in \mathfrak{a}.$$

However, whether we can get something generalizing the full statement of Lemma 3.3.1 seems to be more subtle. ◊

COROLLARY 3.3.3. *If* $\mathsf{M} = U_{r,n}$, *then* $\mathfrak{p}_{[n],\emptyset}$, $\mathfrak{p}_{\emptyset,[n]}$ *and* $\mathfrak{p}_{[n],[n]}$ *are the only associated primes of* $S/\mathfrak{a}$.

PROOF. The only cyclic flats of M are $[n]$ and $\emptyset$, so by Theorem 3.2.4(b), the only minimal primes of $S/\mathfrak{a}$ are $\mathfrak{p}_{[n],\emptyset}$ and $\mathfrak{p}_{\emptyset,[n]}$. It therefore remains to show that $\mathfrak{m} = \mathfrak{p}_{[n],[n]}$ is the only other associated prime.

Let $\mathfrak{p} = \mathfrak{p}_{F,G}$ be an associated prime of $S/\mathfrak{a}$, and let $h \in S/\mathfrak{a}$ be homogeneous with $\operatorname{ann}_S(h) = \mathfrak{p}$. Let $u = \deg(h)$. If $u \in (1,1) + \mathbb{N}^2$, then by Lemma 3.3.1, $\mathfrak{p}_{[n],\emptyset}^{n-r} h = 0$ and $\mathfrak{p}_{\emptyset,[n]}^{r} h = 0$, which means that $\mathfrak{p}_{[n],\emptyset}$ and $\mathfrak{p}_{\emptyset,[n]}$ are both contained in $\mathfrak{p}$. Hence, $\mathfrak{m}$ is contained in and therefore equal to $\mathfrak{p}$.

Next, suppose $u \in \mathbb{N} \times \{0\}$. Since $\mathfrak{a}_{(\cdot,0)} = 0$, h is not killed by any f_i. Therefore, $F = \emptyset$. Hence, by Theorem 3.2.4(b), $\mathfrak{p} = \mathfrak{p}_{\emptyset,[n]}$. Similarly, if $u \in \{0\} \times \mathbb{N}$, then $\mathfrak{p} = \mathfrak{p}_{[n],\emptyset}$. □

3.4. Arrangements with non-isomorphic modules of derivations. At first blush, there appears to be no particular reason to expect—or even hope!—that two different arrangements in the same ambient space will have isomorphic modules of derivations, and indeed they do not in general, as was shown by Ziegler in [**Zie89**, Example 8.7]. But then one encounters the class of free arrangements, all of which have (by definition) isomorphic modules of derivations. Maybe, one hopes, this phenomenon generalizes, at least within isomorphism classes of arrangements! Alas, as with so many properties of free arrangements, as soon as we move far enough away from freeness, we lose: In Theorem 3.4.1, we give a recipe for producing combinatorially equivalent arrangements with non-isomorphic modules of derivations. Note that this recipe does not produce Ziegler's example [**Zie89**, Example 8.7]—in fact, as we explain in Remark 3.4.2, this recipe is only useful when the rank is at least 4. It also fails to offer a method to disprove Terao's conjecture: in Theorem 3.5.1, we will see that the hypothesis is never satisfied for free arrangements.

THEOREM 3.4.1. *Let $f, f' \colon W \hookrightarrow \Bbbk^n$ be realizations of a loopless matroid* M, *and let $\mathcal{A}$ and $\mathcal{A}'$ be the induced arrangements in W. If there exists a minimal non-empty cyclic flat $F \in \mathcal{Z}(\mathsf{M})$ of rank at least 3 such that*

$$\{w \in W : f_i(w) = 0 \text{ for all } i \in F\} \neq \{w \in W : f'_i(w) = 0 \text{ for all } i \in F\},$$

then $\mathrm{Der}(\mathcal{A})$ *is not isomorphic to* $\mathrm{Der}(\mathcal{A}')$ *as R-modules.*

PROOF. By Corollary 3.2.12(b), the ideal

$$P = (f_i : i \in F)$$

is an associated prime of $(S/\mathfrak{a})_{(\cdot,1)}$, and by assumption it is not an associated prime of $(S'/\mathfrak{a}')_{(\cdot,1)}$. Therefore, applying [**EHV92**, Th. 1.1(1)], we find that P is an associated prime of $\mathrm{Ext}_R^{\mathrm{rk}_{\mathsf{M}}(F)}((S/\mathfrak{a})_{(\cdot,1)}, R)$ but not of $\mathrm{Ext}_R^{\mathrm{rk}_{\mathsf{M}}(F)}((S'/\mathfrak{a}')_{(\cdot,1)}, R)$. By (3), $\mathrm{Der}(\mathcal{A})$ is a second syzygy of $(S/\mathfrak{a})_{(\cdot,1)}$. Therefore, since $\mathrm{rk}_{\mathsf{M}}(F) \geq 3$ by assumption,

$$\mathrm{Ext}_R^{\mathrm{rk}_{\mathsf{M}}(F)-2}(\mathrm{Der}(\mathcal{A}), R) \cong \mathrm{Ext}_R^{\mathrm{rk}_{\mathsf{M}}(F)}((S/\mathfrak{a})_{(\cdot,1)}, R).$$

The same statement is true for $\mathcal{A}'$. Thus, P is an associated prime of $\mathrm{Ext}_R^{\mathrm{rk}_{\mathsf{M}}(F)-2}(\mathrm{Der}(\mathcal{A}), R)$ but not of $\mathrm{Ext}_R^{\mathrm{rk}_{\mathsf{M}}(F)-2}(\mathrm{Der}(\mathcal{A}'), R)$. Hence, $\mathrm{Der}(\mathcal{A})$ is not isomorphic to $\mathrm{Der}(\mathcal{A}')$. □

REMARK 3.4.2. If the rank of M is equal to 3, then the only cyclic flat of rank at least 3 is $[n]$. But the flat of the arrangement cut out by $[n]$ is the origin irrespective of what realization of M we choose. ◊

EXAMPLE 3.4.3. Consider the arrangement $\mathcal{A}$ in $\Bbbk^4$ given by the columns of the matrix

$$A = \begin{pmatrix} 0 & 1 & 0 & 0 & 1 \\ 0 & 0 & 1 & 0 & 1 \\ 0 & 0 & 0 & 1 & 0 \\ 1 & 1 & 1 & 1 & 1 \end{pmatrix}.$$

The matroid of this arrangement is $\mathsf{M} = U_{1,\{4\}} \oplus U_{3,\{1,2,3,5\}}$. The only non-empty cyclic flat of M is 1235, which has rank 3. Choose a $P \in \mathrm{GL}(\Bbbk^4)$ which does not fix $V(f_1, f_2, f_3, f_5)$. Then the columns of the matrix PA give an arrangement $\mathcal{A}'$ combinatorially equivalent to $\mathcal{A}$, and the flats of $\mathcal{A}$ and $\mathcal{A}'$ corresponding to the flat 1235 of M are distinct. Therefore, by Theorem 3.4.1, $\mathrm{Der}(\mathcal{A})$ is not isomorphic to $\mathrm{Der}(\mathcal{A}')$.

On the other hand, any arrangement $\mathcal{A}$ realizing M is a cone over a rank-3 generic arrangement, so one can easily show that $\mathrm{Der}(\mathcal{A})$ has a minimal free resolution of the form

$$0 \longrightarrow R(-3) \longrightarrow R(-2)^3 \oplus R(-1)^2 \longrightarrow \mathrm{Der}(\mathcal{A}) \longrightarrow 0,$$

using the main result of Rose and Terao [**RT91**, Cor. 3.4.2]. This invites the question of whether one can also generate pairs of combinatorially equivalent arrangements $\mathcal{A}$, $\mathcal{A}'$ for which $\mathrm{Der}(\mathcal{A})$ and $\mathrm{Der}(\mathcal{A}')$ are not only non-isomorphic, but also have differing Betti numbers. ◊

REMARK 3.4.4. The reader may note that, though in the example above $\mathrm{Der}(\mathcal{A}) \not\cong \mathrm{Der}(\mathcal{A}')$ as R-modules, there is an automorphism $\phi\colon R \to R$ induced by the linear transformation P which sends one module of derivations to the other. Indeed, this will be the case whenever realizations $f, f'\colon W \hookrightarrow \Bbbk^n$ differ by a linear automorphism of W. This raises the question of how to recognize pairs of realizations $\mathcal{A}$, $\mathcal{A}'$ for which both $\mathrm{Der}(\mathcal{A}) \not\cong \mathrm{Der}(\mathcal{A}')$ and no ring automorphism of R takes one to the other. ♢

REMARK 3.4.5. Example 3.4.3 generalizes: Suppose M has a non-empty cyclic flat $F \neq [n]$ of rank at least 3, and suppose $\mathcal{A}$ is given by the columns of a matrix A. If there exists a $P \in \mathrm{GL}(W)$ which does not fix the flat of $\mathcal{A}$ corresponding to F (e.g. if $\Bbbk$ is infinite), then the arrangement $\mathcal{A}'$ given by the columns of PA has matroid M, and by Theorem 3.4.1, $\mathrm{Der}(\mathcal{A}) \not\cong \mathrm{Der}(\mathcal{A}')$. ♢

3.5. Applications to free arrangements. Various conditions on hyperplane arrangements have long been known which are necessary for $\mathrm{Der}(\mathcal{A})$ to be free. In this section, we note that the support of the ideal of pairs (Section 3.2) gives a geometric explanation of a result of Kung and Schenck [**KS06**] which, in turn, generalized a result of Ziegler [**Zie89**].

THEOREM 3.5.1 (Cor. 2.3, [**KS06**]). *Let $\mathcal{A}$ be a non-empty arrangement. Then*

$$\operatorname{pdim}_R \mathrm{Der}(\mathcal{A}) \geq \max\bigl\{\mathrm{rk}_{\mathsf{M}} F - 2 : F \in \min\bigl(\mathcal{Z}(\mathsf{M}) \setminus \{\emptyset\}\bigr)\bigr\}.$$

PROOF. Assume that M has a minimal non-empty cyclic flat F of rank $c \geq 3$. We are going to show that $\operatorname{pdim}_R \mathrm{Der}(\mathcal{A}) \geq c - 2$.

Let $M = (S/\mathfrak{a})_{(\cdot,1)}$. According to (4), $\mathrm{Der}(\mathcal{A})(-1)$ is a second syzygy of M. Therefore, because $c \geq 3$,

$$\mathrm{Ext}_R^{c-2}(\mathrm{Der}(A)(-1), R) \cong \mathrm{Ext}_R^c(M, R).$$

It therefore suffices to prove that $\mathrm{Ext}_R^c(M, R) \neq 0$.

By Corollary 3.2.12(b), P_F is an associated (in fact minimal) prime of M of codimension c. So, [**EHV92**, Th. 1.1(1)] guarantees that P_F is also an associated prime of $\mathrm{Ext}_R^c(M, R)$. Thus, the module $\mathrm{Ext}_R^c(M, R) \neq 0$, as claimed. □

This result is stated slightly differently in the original. To translate, we note that F is a minimal cyclic flat of M of rank c if and only if $\mathsf{M}|F$ is a uniform matroid. Provided $c \geq 2$, this is to say that $\mathcal{A}_F$ is a generic arrangement.

REMARK 3.5.2. Our proof of the the bound in Theorem 3.5.1 used only our knowledge of the minimal primes of $(S/\mathfrak{a})_{(\cdot,1)}$. However, by Corollary 3.2.9(a), the same argument applied to all the associated primes shows that

$$\operatorname{pdim}_R \mathrm{Der}(\mathcal{A}) \geq \max\{\mathrm{rk}_{\mathsf{M}} F - 2 : P_F \text{ is an associated prime of } (S/\mathfrak{a})_{(\cdot,1)}\}.$$

This gives another motivation for Question 3.2.16.

Also using Theorem 3.5.1, one reproves the (rather trivial) fact that if $\mathcal{A}$ is free, then every minimal non-empty cyclic flat has rank at most 2. Note that the converse is false: All the minimal non-empty cyclic flats of the arrangement in Example 2.4.3 have rank 2, yet the arrangement is not free (in fact it is not even tame). ♢

If $\mathcal{A}$ is a free simple arrangement, then, its minimal, non-empty cyclic flats have rank exactly 2. Since the Boolean arrangement is the only arrangement with no cyclic flats, this implies an early result of Ziegler:

COROLLARY 3.5.3 (Cor. 7.6, [**Zie89**]). *Let $\mathcal{A}$ be a simple arrangement. If $\mathcal{A}$ is free and all rank-2 flats have size 2, then $\mathcal{A}$ is the Boolean arrangement.*

3.6. The case that $\mathfrak{a}$ is linear type. Recall that an ideal I in a Noetherian ring is *linear type* if the natural map from the symmetric algebra $\mathrm{Sym}(I)$ of I to the Rees algebra $\mathcal{R}(I)$ of I is an isomorphism (see, e.g., [**Hun80**, §2]). Unravelling the definitions involved, one gets Proposition 3.6.1 below.

To state the proposition, we will need the following two facts (both of which can be found in [**Hun80**, §2]): The kernel of the natural map $S[a] \to \mathcal{R}(\mathfrak{a})$ is

$$\mathcal{J} = \{h(a_1, \ldots, a_n) : h(f_1 g_1, \ldots, f_n g_n) = 0\}.$$

The kernel of the natural map $S[a] \to \mathrm{Sym}(\mathfrak{a})$ is the $S[a]$-ideal $\mathcal{L}$ generated by the elements in $\mathcal{J}$ of A-degree 1.

PROPOSITION 3.6.1. (a) *$I_{\mathfrak{X}} = \mathcal{J}_{(\cdot,1;\cdot)}$, i.e. $I_{\mathfrak{X}}$ is the part of $\mathcal{J}$ with $R^\perp$-degree 1.*
(b) *$I_{\log} = \mathcal{L}_{(\cdot,1;\cdot)}$, i.e. $I_{\log}$ is the part of $\mathcal{L}$ with $R^\perp$-degree 1.*
(c) *If $\mathfrak{a}$ is linear type, then $I_{\log} = I_{\mathfrak{X}}$.*

PROOF. (a) This follows immediately from the definition of $I_{\mathfrak{X}}$.

(b) By Proposition 2.4.2, the A-degree 1 part of $I_{\log}$ agrees with that of $I_{\mathfrak{X}}$. Therefore,

$$\mathcal{L}_{(\cdot,1;1)} = \mathcal{J}_{(\cdot,1;1)} = (I_{\mathfrak{X}})_{(\cdot;1)} = (I_{\log})_{(\cdot;1)}.$$

Now use that, by definition, $I_{\log}$ is generated in A-degree 1.

(c) Since $\mathfrak{a}$ is linear type, $\mathcal{J} = \mathcal{L}$. Now use parts (a) and (b). □

For tame arrangements, $I_{\log} = I_{\mathfrak{X}}$ ([**CDFV11**, Cor. 3.8]). By Proposition 3.6.1(c), this is also true if $\mathfrak{a}$ is linear type. So we have implications

tame $\Longrightarrow$ $I_{\log} = I_{\mathfrak{X}}$ $\Longleftarrow$ $\mathfrak{a}$ is linear type

It is therefore natural to ask the following questions:

QUESTION 3.6.2.

(a) *Is the converse of Proposition* 3.6.1(c) *true?*
(b) *Is the converse of* [**CDFV11**, Cor. 3.8] *true?*
(c) *Does tameness imply that $\mathfrak{a}$ is linear type? What about the converse?*

Acknowledgments

The authors would like to thank Takuro Abe, Joseph Bonin, and Uli Walther for some helpful explanations.

References

[ADH23] F. Ardila, G. Denham, and J. Huh, *Lagrangian geometry of matroids*, J. Amer. Math. Soc. **36** (2023), no. 3, 727–794, DOI 10.1090/jams/1009. MR4583774 ↑60, 70

[CDFV11] D. Cohen, G. Denham, M. Falk, and A. Varchenko, *Critical points and resonance of hyperplane arrangements*, Canad. J. Math. **63** (2011), no. 5, 1038–1057, DOI 10.4153/CJM-2011-028-8. MR2866070 ↑60, 62, 63, 64, 75

[CHKS06] F. Catanese, S. Hoşten, A. Khetan, and B. Sturmfels, *The maximum likelihood degree*, Amer. J. Math. **128** (2006), no. 3, 671–697. MR2230921 ↑60

[Den14] G. Denham, *Toric and tropical compactifications of hyperplane complements* (English, with English and French summaries), Ann. Fac. Sci. Toulouse Math. (6) **23** (2014), no. 2, 297–333, DOI 10.5802/afst.1408. MR3205595 ↑60

[DGS12] G. Denham, M. Garrousian, and M. Schulze, *A geometric deletion-restriction formula*, Adv. Math. **230** (2012), no. 4-6, 1979–1994, DOI 10.1016/j.aim.2012.04.003. MR2927361 ↑62

[DS04] H. Derksen and J. Sidman, *Castelnuovo-Mumford regularity by approximation*, Adv. Math. **188** (2004), no. 1, 104–123, DOI 10.1016/j.aim.2003.10.001. MR2084776 ↑60, 68, 69

[DS12] G. Denham and M. Schulze, *Complexes, duality and Chern classes of logarithmic forms along hyperplane arrangements*, Arrangements of hyperplanes—Sapporo 2009, Adv. Stud. Pure Math., vol. 62, Math. Soc. Japan, Tokyo, 2012, pp. 27–57, DOI 10.2969/aspm/06210027. MR2933791 ↑64

[EHV92] D. Eisenbud, C. Huneke, and W. Vasconcelos, *Direct methods for primary decomposition*, Invent. Math. **110** (1992), no. 2, 207–235, DOI 10.1007/BF01231331. MR1185582 ↑73, 74

[FHG+21] R. Freij-Hollanti, M. Grezet, C. Hollanti, and T. Westerbäck, *Cyclic flats of binary matroids*, Adv. in Appl. Math. **127** (2021), Paper No. 102165, 47, DOI 10.1016/j.aam.2021.102165. MR4207227 ↑68

[GS] D. Grayson and M. Stillman, *Macaulay2—a software system for algebraic geometry and commutative algebra*, available at `http://www.math.uiuc.edu/Macaulay2`. 64, 69, 70, 71

[HS14] J. Huh and B. Sturmfels, *Likelihood geometry*, Combinatorial algebraic geometry, Lecture Notes in Math., vol. 2108, Springer, Cham, 2014, pp. 63–117, DOI 10.1007/978-3-319-04870-3_3. MR3329087 ↑60, 63

[Huh13] J. Huh, *The maximum likelihood degree of a very affine variety*, Compos. Math. **149** (2013), no. 8, 1245–1266, DOI 10.1112/S0010437X13007057. MR3103064 ↑62

[Hun80] C. Huneke, *On the symmetric and Rees algebra of an ideal generated by a d-sequence*, J. Algebra **62** (1980), no. 2, 268–275, DOI 10.1016/0021-8693(80)90179-9. MR563225 ↑75

[KS06] J. P. S. Kung and H. Schenck, *Derivation modules of orthogonal duals of hyperplane arrangements*, J. Algebraic Combin. **24** (2006), no. 3, 253–262, DOI 10.1007/s10801-006-0023-6. MR2260017 ↑61, 74

[OT92] P. Orlik and H. Terao, *Arrangements of hyperplanes*, Grundlehren der mathematischen Wissenschaften [Fundamental Principles of Mathematical Sciences], vol. 300, Springer-Verlag, Berlin, 1992, DOI 10.1007/978-3-662-02772-1. MR1217488 ↑63, 66

[OT95a] P. Orlik and H. Terao, *Arrangements and Milnor fibers*, Math. Ann. **301** (1995), no. 2, 211–235, DOI 10.1007/BF01446627. MR1314585 ↑64

[OT95b] P. Orlik and H. Terao, *The number of critical points of a product of powers of linear functions*, Invent. Math. **120** (1995), no. 1, 1–14, DOI 10.1007/BF01241120. MR1323980 ↑60

[Oxl11] J. Oxley, *Matroid theory*, 2nd ed., Oxford Graduate Texts in Mathematics, vol. 21, Oxford University Press, Oxford, 2011, DOI 10.1093/acprof:oso/9780198566946.001.0001. MR2849819 ↑61, 68, 70

[RT91] L. L. Rose and H. Terao, *A free resolution of the module of logarithmic forms of a generic arrangement*, J. Algebra **136** (1991), no. 2, 376–400, DOI 10.1016/0021-8693(91)90052-A. MR1089305 ↑73

[Var95] A. Varchenko, *Critical points of the product of powers of linear functions and families of bases of singular vectors*, Compositio Math. **97** (1995), no. 3, 385–401. MR1353281 ↑60

[Zie89] G. M. Ziegler, *Combinatorial construction of logarithmic differential forms*, Adv. Math. **76** (1989), no. 1, 116–154, DOI 10.1016/0001-8708(89)90045-5. MR1004488 ↑61, 72, 74, 75

Department of Mathematics, University of Western Ontario, London, Ontario N6A 5B7, Canada

Email address: gdenham@uwo.ca

URL: http://gdenham.math.uwo.ca/

Department of Mathematics, University of Western Ontario, London, Ontario N6A 5B7, Canada

Email address: asteine@uwo.ca

URL: https://sites.google.com/view/avi-steiner/

Contemporary Mathematics
Volume **790**, 2023
https://doi.org/10.1090/conm/790/15859

Shift of argument algebras and de Concini–Procesi spaces

Iva Halacheva

Abstract. In this expository article, we recall the construction and properties of the de Concini–Procesi wonderful compactification $\overline{\mathcal{M}}_{\mathfrak{g}}$ associated to the root hyperplane configuration of a semisimple Lie algebra $\mathfrak{g}$. We then describe the structure of the family of shift of argument algebras, a family of maximal Poisson-commutative subalgebras of $S(\mathfrak{g})$ parametrized by elements $\mu \in \mathbb{P}(\mathfrak{h}^{\mathrm{reg}})$, and discuss how one can compactify it to a family parametrized by $\overline{\mathcal{M}}_{\mathfrak{g}}$, as well as lift it to $\mathcal{U}(\mathfrak{g})$. When considering the real locus $\overline{\mathcal{M}}_{\mathfrak{g}}(\mathbb{R})$, any shift of argument algebra corresponding to such a parameter acts with simple spectrum on a given highest weight irreducible $\mathfrak{g}$-representation $V(\lambda)$. We recall that this produces a covering of $\overline{\mathcal{M}}_{\mathfrak{g}}(\mathbb{R})$ such that the monodromy action on the fibers coincides with the cactus group action $C_{\mathfrak{g}}$ on the crystal $B(\lambda)$.

1. Introduction

The *de Concini–Procesi model* gives a way to build an algebraic variety with intricate geometric and combinatorial properties starting from certain subspace configurations called *building sets* [**5**]. In particular, this construction can be considered in the case of the hyperplane arrangement associated to the root system of a semisimple Lie algebra $\mathfrak{g}$. The resulting models and their real loci are revealed to have an interesting operadic and topological structure. In the case of the Lie algebra $\mathfrak{sl}_n$, the de Concini–Procesi model coincides with the Deligne–Mumford moduli space of stable genus 0 curves with $n+1$ marked points [**7**]. These geometric models turn out to have further connections to Lie theory, in the setting of a family of subalgebras of the symmetric algebra $S(\mathfrak{g})$ called *shift of argument algebras.*

Originally introduced by Mishchenko and Fomenko [**10**], shift of argument algebras are a family of Poisson-commutative subalgebras of $S(\mathfrak{g})$ parametrized by $\mathfrak{g}$. For a regular element, $\mu \in \mathfrak{g}^{\mathrm{reg}}$, the corresponding subalgebra A_μ is maximal Poisson-commutative in $S(\mathfrak{g})$ ([**35**], [**25**]). Moreover, these algebras remain unchanged under the scaling of the parameter. We will restrict our attention to the family parametrized by projectivized regular Cartan elements, i.e. by $\mathbb{P}(\mathfrak{h}^{\mathrm{reg}})$. Recent work (see [**14**]) has shown that there is a way to take the closure of this family, by considering limits in the appropriate Grassmannians, to obtain a smooth algebraic variety isomorphic to the corresponding de Concini–Procesi model $\overline{\mathcal{M}}_{\mathfrak{g}}$.

2020 *Mathematics Subject Classification.* Primary 54C40, 14E20; Secondary 46E25, 20C20.
Key words and phrases. Shift of argument algebras, wonderful model, cactus group.

This expository article is organized as follows: Section 2 recalls the general construction of the *minimal* de Concini–Procesi model for a root hyperplane arrangement, as well as its real locus $\overline{\mathcal{M}}_{\mathfrak{g}}(\mathbb{R})$. Section 3 describes this space and some of its properties in the case of $\mathfrak{sl}_n$. Sections 4 and 5 outline the general structure of shift of argument algebras, and the isomorphism between the closure of that family and the de Concini–Procesi model, as well as their lifts to the universal enveloping algebra $\mathcal{U}(\mathfrak{g})$, and their operadic nature. In Section 6, we discuss the covering space $E(\lambda) \to \overline{\mathcal{M}}_{\mathfrak{g}}(\mathbb{R})$ obtained by considering the action of shift of argument algebras on a fixed highest weight irreducible $\mathfrak{g}$-representation $V(\lambda)$, as well as the resulting monodromy action which recovers the cactus group action on crystals.

2. The de Concini–Procesi moduli space

Let $\mathfrak{g}$ denote a complex semisimple Lie algebra of rank $\ell := \operatorname{rk}\mathfrak{g}$, Cartan subalgebra $\mathfrak{h}$, Dynkin diagram I, Weyl group $W = W_I$, sets of positive and all roots $\Delta_+ \subseteq \Delta \subseteq \mathfrak{h}^*$ respectively, simple roots $\{\alpha_i\}_{i\in I}$, and root hyperplanes in $\mathfrak{h}$ denoted $H_\alpha = \{h \in \mathfrak{h} : \alpha(h) = 0\}$ for $\alpha \in \Delta$. We will denote the regular elements of the Cartan by:

$$\mathfrak{h}^{\text{reg}} = \mathfrak{h} \setminus \bigcup_{\alpha\in\Delta} H_\alpha = \{\mu \in \mathfrak{h} \;:\; \alpha_i(\mu) \neq 0 \,\forall\, i \in I\}$$

i.e. the Cartan elements with minimal centralizer dimension, or equivalently such that no root of $\mathfrak{g}$ evaluated on such an element vanishes.

2.1. Minimal building set. Consider the finite set $\mathcal{C}$ whose elements are the spaces spanned by subsets of the roots of $\mathfrak{g}$,

$$\mathcal{C} = \{V \subseteq \mathfrak{h}^* : V = \operatorname{Span}_{\mathbb{C}}\{\alpha \mid \alpha \in \Delta'\} \text{ for some subset } \Delta' \subseteq \Delta\}.$$

An expression of the form $V = V_1 \oplus V_2 \oplus \cdots \oplus V_k$ is called a **decomposition of** V if $V_1, \ldots, V_k$ are subspaces of $\mathfrak{h}^*$ and for any root $\alpha \in V \cap \Delta$, we have that $\alpha \in V_i$ for some i. Every subspace $V \in \mathcal{C}$ has a unique decomposition into **indecomposables**, i.e. elements which admit only the trivial decomposition ([**5**, Section 2.1]). The **minimal building set** $\mathcal{D}$ associated to Δ is the collection of indecomposable elements of $\mathcal{C}$, or in other words any subspace of $\mathfrak{h}^*$ which is the span of the roots of an irreducible root subsystem of Δ. Every element of the minimal building set $\mathcal{D}$ is of the form $w(V_J)$ for some $V_J = \operatorname{Span}_{\mathbb{C}}\{\alpha : \alpha \in J\}$, where $J \subseteq I$ is a non-empty connected subdiagram, $w \in W$, and the Weyl group action comes from the action of W on $\Delta \subseteq \mathfrak{h}^*$.

EXAMPLE 2.1. In the case $\mathfrak{g} = \mathfrak{sl}_4$, the simple roots can be enumerated as $\{\alpha_1, \alpha_2, \alpha_3\}$, and the collection of all roots is $\Delta = \{\pm\alpha_1, \pm\alpha_2, \pm\alpha_3, \pm(\alpha_1 + \alpha_2), \pm(\alpha_2 + \alpha_3), \pm(\alpha_1 + \alpha_2 + \alpha_3)\}$. The corresponding minimal building set is:

$$\mathcal{D} = \{V_1, V_2, V_3, V_{12}, V_{23}, V_{123}, V_{1,2}, V_{2,3}, V_{12,3}, V_{1,23}\}$$

where any subscript of the form i, ij or ijk indicates that the span of the root $\alpha_i, \alpha_i + \alpha_j$, or $\alpha_i + \alpha_j + \alpha_k$ respectively is included. For instance, $V_{1,2} = \operatorname{Span}_{\mathbb{C}}\{\alpha_1, \alpha_2\}$, and $V_{1,23} = \operatorname{Span}_{\mathbb{C}}\{\alpha_1, \alpha_2 + \alpha_3\}$. Note that $V_{1,3}$ does not belong to the minimal building set as it has a nontrivial decomposition, $V_{1,3} = V_1 \oplus V_3$.

2.2. Construction of the moduli space. Given a subspace of the form $V = \mathrm{Span}_{\mathbb{C}}\{\alpha \mid \alpha \in \Delta' \subseteq \Delta\} \subseteq \mathfrak{h}^*$, its annihilator is $V^\perp = \bigcap_{\alpha\in\Delta'} H_\alpha \subseteq \mathfrak{h}$. We consider the map

$$\psi : \mathcal{M}_{\mathfrak{g}} := \mathbb{P}(\mathfrak{h}^{\mathrm{reg}}) \to \prod_{V\in\mathcal{D}} \mathbb{P}(\mathfrak{h}/V^\perp)$$

whose components are obtained from the restrictions of the projection maps for each $V \in \mathcal{D}$, $\mathfrak{h} \setminus V^\perp \to \mathbb{P}(\mathfrak{h}/V^\perp)$.

DEFINITION 2.2 ([**5**, Definition 1.1]). The de Concini–Procesi **wonderful model** $\overline{\mathcal{M}}_{\mathfrak{g}}$ associated to $\mathfrak{g}$ is a smooth compactification of $\mathcal{M}_{\mathfrak{g}}$ defined as the closure of the image of the map ψ:

$$\overline{\mathcal{M}}_{\mathfrak{g}} := \overline{\mathrm{Im}\,\psi} \subseteq \prod_{V\in\mathcal{D}} \mathbb{P}(\mathfrak{h}/V^\perp).$$

Note that if $\mathfrak{g} = \mathfrak{g}_1 \oplus \mathfrak{g}_2$, then $\Delta = \Delta_1 \sqcup \Delta_2$, $\mathcal{D} = \mathcal{D}_1 \sqcup \mathcal{D}_2$ and $\overline{\mathcal{M}}_{\mathfrak{g}} \cong \overline{\mathcal{M}}_{\mathfrak{g}_1} \times \overline{\mathcal{M}}_{\mathfrak{g}_2}$, so it suffices to consider the case where $\mathfrak{g}$ is a simple Lie algebra. We assume that $\mathfrak{g}$ is simple from now on, unless otherwise specified. In that case, we also have that ψ is injective, as $\mathfrak{h}^* \in \mathcal{D}$, so the target space of ψ includes the factor $\mathbb{P}(\mathfrak{h})$.

$\overline{\mathcal{M}}_{\mathfrak{g}}$ can equivalently be defined as an iterated blow-up along successive intersections of root hyperplanes [**5**, Section 3]. It is a smooth projective variety, contains $\mathcal{M}_{\mathfrak{g}}$ as a dense open subset, and $\overline{\mathcal{M}}_{\mathfrak{g}} \setminus \mathcal{M}_{\mathfrak{g}}$ is a smooth normal crossing divisor. A point in $\overline{\mathcal{M}}_{\mathfrak{g}}$ can be thought of as a tuple $(\ell_V)_{V\in\mathcal{D}}$ where ℓ_V is a line in $\mathfrak{h}/V^\perp$ or equivalently a subspace of $\mathfrak{h}$ that contains $V^\perp$ as a hyperplane.

As mentioned above, since we assume that $\mathfrak{g}$ is simple, the target space of ψ includes the factor $\mathbb{P}(\mathfrak{h})$, and since $\mathfrak{h}^{\mathrm{reg}}$ is dense in $\mathfrak{h}$, we get a surjective map

$$p : \overline{\mathcal{M}}_{\mathfrak{g}} \to \mathbb{P}(\mathfrak{h})$$

coming from the projection onto that factor. The map p is an isomorphism over the locus $\mathbb{P}(\mathfrak{h}^{\mathrm{reg}})$. Indeed if $\mu \in \mathfrak{h}^{\mathrm{reg}}$, then $\mu \notin V^\perp$ for any $V \in \mathcal{D}$, so $p^{-1}([\mu]) = (\mathbb{C}\mu \oplus V^\perp)_{V\in\mathcal{D}}$.

Moreover, the following operadic property holds for $\overline{\mathcal{M}}_{\mathfrak{g}}$ when considering the fibers of p (see e.g. [**14**]), which will be useful in Section 5.3.

LEMMA 2.3. *For any $\mu \in \mathfrak{h}$, consider its centralizer in $\mathfrak{g}$, denoted $\mathfrak{g}_\mu$. Then the root system of $\mathfrak{g}'_\mu = [\mathfrak{g}_\mu, \mathfrak{g}_\mu]$ is isomorphic to $\Delta_\mu = \{\alpha \in \Delta \ : \ \alpha(\mu) = 0\}$, and*

$$p^{-1}([\mu]) \cong \overline{\mathcal{M}}_{\mathfrak{g}'_\mu}.$$

EXAMPLE 2.4. Let $\mathfrak{g} = \mathfrak{sl}_6$ and $\mu = (1, 1, 1, -1, -1, -1)$. Then $\mathfrak{g}'_\mu = [\mathfrak{g}_\mu, \mathfrak{g}_\mu] = \mathfrak{sl}_3 \oplus \mathfrak{sl}_3$ so $p^{-1}([\mu]) \cong \overline{\mathcal{M}}_{\mathfrak{sl}_3\oplus\mathfrak{sl}_3} \cong \overline{\mathcal{M}}_{\mathfrak{sl}_3} \times \overline{\mathcal{M}}_{\mathfrak{sl}_3} \cong \mathbb{P}^1 \times \mathbb{P}^1$ (see also Section 3).

2.3. The real locus and the cactus group. We recall some properties of the real locus of the de Concini–Procesi moduli space, $\overline{\mathcal{M}}_{\mathfrak{g}}(\mathbb{R})$, which we can consider since the root system of $\mathfrak{g}$ is defined over $\mathbb{R}$ and therefore so is the resulting variety $\overline{\mathcal{M}}_{\mathfrak{g}}$. This real locus turns out to be quite different topologically from $\overline{\mathcal{M}}_{\mathfrak{g}}$. Below, we summarize some of these differences and discuss its fundamental group. For the following definition, see for instance [**15**], [**13**], [**23**].

DEFINITION 2.5. The **cactus group** $C_{\mathfrak{g}}$ associated to the Lie algebra $\mathfrak{g}$ has generators c_J indexed by connected Dynkin subdiagrams $J \subseteq I$, and three types of relations:

(1) $c_J^2 = 1$,
(2) $c_J c_K = c_K c_{\theta_K(J)}$ for all $J \subseteq K$,
(3) $c_J c_K = c_K c_J$ for all J, K such that $J \cup K$ is disconnected.

In this definition, θ_K is the Dynkin diagram automorphism of the subdiagram K defined as follows: let w_0^K denote the longest element in the parabolic Weyl group W_K of the Levi subalgebra corresponding to K (and $w_0 = w_0^I$ is the longest element of W). Then for any $k \in K$, we have $\alpha_{\theta_K(k)} = -w_0^K \cdot \alpha_k$.

There is a short exact sequence $1 \to \mathrm{PC}_{\mathfrak{g}} \xrightarrow{\iota} \mathrm{C}_{\mathfrak{g}} \xrightarrow{\kappa} W \to 1$ where for any connected $J \subseteq I$, $\kappa(c_J) = w_0^J$ with the notation as above, and ι is the inclusion of $\mathrm{PC}_{\mathfrak{g}} = \operatorname{Ker} \kappa$. The group $\mathrm{PC}_{\mathfrak{g}}$ is called the **pure cactus group**. It follows from work of Davis–Januszkiewicz–Scott [**4**] that $\mathrm{PC}_{\mathfrak{g}}$ is the fundamental group of $\overline{\mathcal{M}}_{\mathfrak{g}}(\mathbb{R})$. More generally, the cactus group $\mathrm{C}_{\mathfrak{g}}$ is the *W-equivariant fundamental group* of $\overline{\mathcal{M}}_{\mathfrak{g}}(\mathbb{R})$ as described below, where $\sim$ denotes equivalence up to homotopy of paths, and $*$ denotes path concatenation. We fix a basepoint $\mu \in \overline{\mathcal{M}}_{\mathfrak{g}}(\mathbb{R})$.

$$\pi_1^W(\overline{\mathcal{M}}_{\mathfrak{g}}(\mathbb{R}), \mu) = \{(\gamma, w) : w \in W, \gamma \text{ is a path from } \mu \text{ to } w\mu\}/\sim$$

$$(\gamma_1, w_1) \cdot (\gamma_2, w_2) = (w_1(\gamma_2) * \gamma_1, w_1 w_2)$$

To summarize:

THEOREM 2.6 ([**4**], Theorem 4.7.2). *Let μ denote a basepoint in $\overline{\mathcal{M}}_{\mathfrak{g}}(\mathbb{R})$, then the pure cactus group and cactus group satisfy:*

$$\mathrm{PC}_{\mathfrak{g}} \cong \pi_1(\overline{\mathcal{M}}_{\mathfrak{g}}(\mathbb{R}), \mu) \qquad\qquad \mathrm{C}_{\mathfrak{g}} \cong \pi_1^W(\overline{\mathcal{M}}_{\mathfrak{g}}(\mathbb{R}), \mu).$$

In fact, $\overline{\mathcal{M}}_{\mathfrak{g}}(\mathbb{R})$ is an Eilenberg–MacLane space as it is $K(\pi, 1)$ for $\mathrm{PC}_{\mathfrak{g}}$ (see [**4**, Corollary 5.9.4]). We return to this point in Section 3 in the case of $\mathfrak{g} = \mathfrak{sl}_n$. The isomorphism in Theorem 2.6 can be constructed as follows (see for instance [**14**, Section 2.13], [**6**, Theorem 3.2]). Consider the subset P_+ of $\overline{\mathcal{M}}_{\mathfrak{g}}(\mathbb{R})$ given by the image of the open dominant Weyl chamber under ψ:

$$P_+ = \psi(\{\mu \in \mathfrak{h}(\mathbb{R}) \ : \ \alpha_i(\mu) > 0 \, \forall \, i \in I\}).$$

Then its closure $\overline{P}_+$ is (homeomorphic to) a convex polytope, and in particular is contractible. So, for any two points $p, q \in \overline{P}_+$ there is a unique path $\gamma_{p,q}$ up to homotopy connecting them in $\overline{P}_+$. We can assume that the basepoint μ is in P_+ and that $w_0(\mu) = \mu$ in $\overline{\mathcal{M}}_{\mathfrak{g}}(\mathbb{R})$.

The isomorphism $\mathrm{C}_{\mathfrak{g}} \cong \pi_1^W(\overline{\mathcal{M}}_{\mathfrak{g}}(\mathbb{R}), \mu)$ in Theorem 2.6 is realized by, for any connected $J \subseteq I$,

$$c_J \mapsto (\gamma_{\mu, w_0^J(\mu)}, w_0^J),$$

where w_0^J is a longest parabolic Weyl group element as before and the path $\gamma_{\mu, w_0^J(\mu)}$ from μ to $w_0^J(\mu)$ can be constructed as a concatenation of two paths:

- Choose any point μ^J on the J-facet of $\overline{P}_+$, i.e.
$$\mu^J \in \overline{P}_+ \cap p^{-1}(\{\nu \in \mathfrak{h}(\mathbb{R}) \ : \ \alpha_k(\nu) = 0 \text{ for } k \in J, \alpha_k(\nu) > 0 \text{ for } k \in I \setminus J\}),$$
and such that $w_0^J(\mu^J) = \mu^J$. Take the (unique up to homotopy) path γ_{μ,μ^J}.
- Take the path $\gamma_{\mu^J, w_0^J(\mu)}$ obtained from traversing the w_0^J–translate of γ_{μ,μ^J} in reverse.
- Finally, take $\gamma_{\mu, w_0^J(\mu)} = \gamma_{\mu^J, w_0^J(\mu)} * \gamma_{\mu,\mu^J}$.

3. The $\mathfrak{sl}_n$ case: the Deligne–Mumford moduli space

In the case $\mathfrak{g} = \mathfrak{sl}_n$, we denote the set of nodes of the Dynkin diagram $I = \{1, 2, \dots, n-1\}$, and simple roots $\alpha_i = \varepsilon_i - \varepsilon_{i+1}, \forall\ i \in I$, as well as all roots $\Delta = \{\varepsilon_i - \varepsilon_j\ :\ i \neq j \in \{1, 2, \dots, n\}\}$,

$$\mathfrak{h} = \mathbb{C}^n/\mathbb{C}(1, \dots, 1), \qquad \mathfrak{h}^* = \{(h_1, \dots, h_n) \in \mathbb{C}^n\ :\ \sum_{i=1}^{n} h_i = 0\}.$$

In this setting, the relevant complex moduli space is the **Deligne–Mumford moduli space** $\overline{\mathrm{M}}_{0,n+1}$ ([**7**]) of stable (possibly singular) genus 0 curves with $n+1$ marked points obtained as a compactification of the quotient $\mathrm{M}_{0,n+1} := ((\mathbb{CP}^1)^{n+1} \setminus \Delta)/\operatorname{PGL}(2)$ by the automorphism group $\operatorname{PGL}(2)$ of $\mathbb{CP}^1$, where

$$\Delta = \{(z_1, \dots, z_{n+1}) \in (\mathbb{P}^1)^{n+1}\ :\ z_i = z_j \text{ for some } i \neq j\}$$

is the thick diagonal. $\overline{\mathrm{M}}_{0,n+1}$ is a smooth projective variety of dimension $n-2$ whose points are given by equivalence classes of a stable curves.

A **stable genus** 0 **curve** C **with** $n+1$ **marked points** is a union of finitely many projective lines $C_1, \dots, C_m$, with $n+1$ labeled distinct points $z_1, z_2, \dots, z_{n+1}$ such that:

- each point z_i belongs to exactly one C_j,
- any two components C_i, C_j either don't intersect or intersect transversely in a single point,
- the component graph with vertices corresponding to the C_i's and edges corresponding to intersection points $C_i \cap C_j$ is a tree,
- the total number of marked or intersection points on each component is at least 3.

An **equivalence between two stable curves** $C \to C'$ with the same number of components and marked points is an isomorphism of algebraic curves which (possibly permutes the components and) maps the i–th marked point of C to the i–th marked point of C'.

Intuitively, in the compactification, as some of the $n+1$ marked points on a given component collide, they form a new component on which the collided points live.

3.1. Operads and connection to de Concini–Procesi's construction.

The Deligne–Mumford space $\overline{\mathrm{M}}_{0,n+1}$ and the de Concini–Procesi space $\overline{\mathcal{M}}_{\mathfrak{sl}_n}$ contain dense open sets which are isomorphic to each other, $\overline{\mathrm{M}}_{0,n+1} \supseteq \mathrm{M}_{0,n+1} \cong \mathcal{M}_{\mathfrak{sl}_n} \subseteq \overline{\mathcal{M}}_{\mathfrak{sl}_n}$. Indeed, for $\overline{\mathrm{M}}_{0,n+1}$ the action of $\operatorname{PGL}(2)$ is transitive, so we can use it to fix the last marked point $z_{n+1} = \infty$. The remaining action is of the Borel subgroup B of upper triangular matrices in $\operatorname{PGL}(2)$, which acts on $\mathbb{C}^n$ by diagonal scaling and translation, i.e. $B \cong \mathbb{C}^\times \ltimes \mathbb{C}$. So, we can set the sum of the remaining n points to be zero, and one obtains the isomorphism by (see also [**14**, Section 3.1]):

$$\mathrm{M}_{0,n+1} = ((\mathbb{CP}^1)^{n+1} \setminus \Delta)/\operatorname{PGL}(2) \cong (\mathbb{C}^n \setminus \Delta)/(\mathbb{C}^\times \ltimes \mathbb{C}) \cong \mathfrak{h}^{\mathrm{reg}}_{\mathfrak{sl}_n}/\mathbb{C}^\times = \mathcal{M}_{\mathfrak{sl}_n}.$$

By [**5**, Section 4.3] and [**18**], this isomorphism can be extended to an isomorphism on the compactifications $\overline{\mathcal{M}}_{\mathfrak{sl}_n} \cong \overline{\mathrm{M}}_{0,n+1}$.

The Deligne–Mumford space $\overline{\mathrm{M}}_{0,n+1}$ has a stratification, where the strata are indexed by rooted trees (with root at $z_{n+1} = \infty$) and n leaves colored by the points $z_1, \dots, z_n$. In particular, the stratum indexed by a tree T_1 is in the closure of the stratum indexed by T_2 precisely when T_2 can be obtained from T_1 by contracting some edges. Furthermore, $\overline{\mathrm{M}}_{0,n+1}$ has an operad structure–one can think of a point in $M(n) := \overline{\mathrm{M}}_{0,n+1}$ as an n–ary operation with inputs the marked points 1 through n and output the marked point $n+1$. Then $M(\bullet)$ forms a topological operad with substitution maps $\overline{\mathrm{M}}_{0,a+1} \circ_i \overline{\mathrm{M}}_{0,b+1} \to \overline{\mathrm{M}}_{0,a+b}$ given by gluing the i–th marked point of a curve with $a+1$ labeled points to the $(b+1)$–st marked point of a curve with $b+1$ labeled points (see [**12**, Section 1.4], [**24**]).

3.2. The real locus $\overline{\mathrm{M}}_{0,n+1}(\mathbb{R})$. As mentioned above, the topology of the real locus $\overline{\mathrm{M}}_{0,n+1}(\mathbb{R})$ turns out to be quite different from that of $\overline{\mathrm{M}}_{0,n+1}$, and has been of interest to researchers for some time ([**8**], [**9**], [**11**], [**12**], [**19**]). Over the real numbers, a stable curve becomes an $(n+1)$–*fruit cactus*, i.e. an algebraic curve over $\mathbb{R}$ with $n+1$ marked points and properties as discussed above. Such a cactus can be visualized as a tree of circles with $n+1$ labeled points on them. For instance, the following cactus represents a point in $\overline{\mathrm{M}}_{0,9}(\mathbb{R})$:

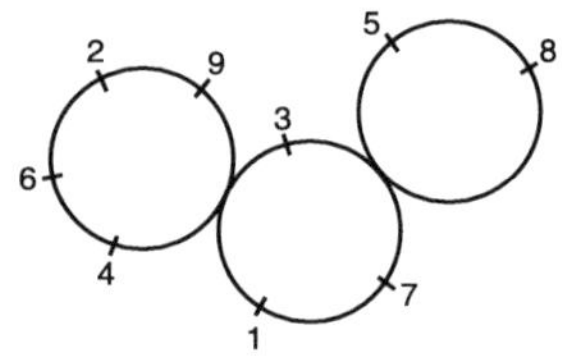

EXAMPLE 3.1. $\overline{\mathrm{M}}_{0,3}(\mathbb{R})$ is a point, $\overline{\mathrm{M}}_{0,4}(\mathbb{R})$ is a circle, and $\overline{\mathrm{M}}_{0,5}(\mathbb{R})$ is the connected sum of 5 real projective planes. See [**8**] for further details on the structure of $\overline{\mathrm{M}}_{0,n}(\mathbb{R})$.

The fundamental group of $\overline{\mathcal{M}}_{\mathfrak{sl}_n}(\mathbb{R}) \cong \overline{\mathrm{M}}_{0,n+1}(\mathbb{R})$ was computed by Davis–Januszkiewicz–Scott [**4**], and the stratification of $\overline{\mathrm{M}}_{0,n+1}(\mathbb{R})$ described by Devadoss and Kapranov gives a cell decomposition, through which one can also compute its Euler characteristic (see [**8**], [**17**]). As in the case of $\overline{\mathrm{M}}_{0,n+1}$, the collection of spaces $\overline{\mathrm{M}}_{0,n+1}(\mathbb{R})$ also form a topological operad compatible with the cell decomposition, described in [**8**] and called the *mosaic operad*, by attaching cacti along their needles. Rains shows in [**26**] that more generally $\overline{\mathcal{M}}_{\mathfrak{g}}(\mathbb{R})$ also has a type of generalized operadic nature, see also Lemma 2.3.

Work of Keel [**18**] computes the integral cohomology ring of the complex moduli space, $H^*(\overline{\mathrm{M}}_{0,n+1}, \mathbb{Z})$, which is commutative and generated by a collection of degree 2 elements indexed by subsets of $\{1, 2, \ldots, n+1\}$ of size between 2 and $n-1$, subject to certain relations (see also [**9**, Section 5.1]). In [**9**], Etingof et al. use Keel's work to construct a convenient basis of monomials in the above generators for $H^*(\overline{\mathrm{M}}_{0,n+1}, \mathbb{Z})$, and determine a presentation and a basis for the algebra $H^*(\overline{\mathrm{M}}_{0,n+1}(\mathbb{R}), \mathbb{Q})$, as well as compute the Betti numbers of the real locus. Several of the conjectures in [**9**] were later resolved by Khoroshkin and Willwacher in [**19**]. As part of their work, they realize the operad of $\overline{\mathrm{M}}_{0,n+1}(\mathbb{R})$ as a homotopy quotient of the operad of associative algebras. Below, we summarize some known results about the topology of $\overline{\mathrm{M}}_{0,n+1}(\mathbb{R})$ as well as the structure of its rational cohomology:

THEOREM 3.2. *The following holds for the real locus of the Deligne–Mumford moduli space, $\overline{\mathrm{M}}_{0,n+1}(\mathbb{R})$, where $n \geq 2$.*

(1) [**4, 8**] *It is a smooth, connected, compact manifold of dimension $n-2$.*

(2) [4] *It is a* $K(\pi,1)$ *space for its fundamental group* $\mathrm{PC}_n := \mathrm{PC}_{\mathfrak{sl}_n}$. *So, in particular* PC_n *is finitely-presented and torsion-free.*
(3) [19] *It is moreover a rational* $K(\pi,1)$ *space, i.e. its* $\mathbb{Q}$*-completion is* $K(\pi,1)$.
(4) [8,11] *Its Euler characteristic can be computed via its cell decomposition, and is given by* $(-1)^{n/2}(n-1)!!(n-3)!!$ *for even* n, *and by* 0 *for odd* n.
(5) [9,26] *For* $n \geq 4$, $\overline{\mathrm{M}}_{0,n+1}(\mathbb{R})$ *is non-orientable.* $H^*(\overline{\mathrm{M}}_{0,n+1}(\mathbb{R}),\mathbb{Z})$ *contains 2–torsion but not 4–torsion, and has no odd torsion.*
(6) [9] *Its Poincaré polynomial is*

$$P_{n+1}(t) = \prod_{0\leq i<(n-2)/2} (1+(n-2-2i)^2 t).$$

(7) [9] $H^*(\overline{\mathrm{M}}_{0,n+1}(\mathbb{R}),\mathbb{Q})$ *is the skew-commutative* $\mathbb{Q}$*–algebra with generators* $h_{i,j,k}$, $1 \leq i,j,k \leq n$, *anti-symmetric in the indices, satisfying two types of quadratic relations:*

$$h_{i,j,k}h_{i,j,l} = 0, \quad h_{i,j,k}h_{k,l,m}+h_{j,k,l}h_{l,m,i}+h_{k,l,m}h_{m,i,j}+h_{l,m,i}h_{i,j,k}+h_{m,i,j}h_{j,k,l} = 0.$$

REMARK 3.3. As pointed out in [9], $\overline{\mathrm{M}}_{0,n+1}(\mathbb{R})$ is topologically more similar to $(\mathbb{C}^n \setminus \Delta) \cong \mathfrak{h}^{\mathrm{reg}}_{\mathfrak{sl}_{n+1}}$ than to $\overline{\mathrm{M}}_{0,n+1}$. In particular, $\mathfrak{h}^{\mathrm{reg}}_{\mathfrak{sl}_{n+1}}$ is also a $K(\pi,1)$ space for its fundamental group, the pure braid group $\mathrm{PBr}_n = \pi_1(\mathfrak{h}^{\mathrm{reg}}_{\mathfrak{sl}_{n+1}},\mu)$. Its Poincaré polynomial also factors nicely as $\prod_{1\leq i\leq n-1}(1+it)$, and its Betti numbers also grow polynomially in n, while $\overline{\mathrm{M}}_{0,n+1}$ is simply connected, does not have a simple factorization of its Poincaré polynomial, and its Betti numbers grow exponentially. Hence, there is an analogy between the pure cactus group $\mathrm{PC}_n = \pi_1(\overline{\mathrm{M}}_{0,n+1}(\mathbb{R}),\mu)$ and the pure braid group $\mathrm{PBr}_n = \pi_1(\mathfrak{h}^{\mathrm{reg}}_{\mathfrak{sl}_{n+1}},\mu)$.

Moreover, there is a map from the cactus group to the prounipotent completion of the braid group,

$$\mathrm{C}_n \to \widehat{\mathrm{Br}_n},$$

achieved using Drinfeld's unitarization construction. Through this construction, one can obtain a cactus commutor R_C from the braid commutor R, by taking $R_\mathrm{C} = R(R^{21}R)^{-1/2}$ (see [15, Section 2.6]). It would be interesting to find out more about this map, for instance by exploring the originally stated conjecture in [9, Section 3.7] that PC_n injects into its prounipotent completion. Khoroshkin and Willwacher reinterpret this map, or more generally a functorial assignment of a coboundary category structure coming from a braided category structure, through the deformation theory of operads as a map from a model of the mosaic operad to a model associated to the little disks operad (see [19, Section 7]).

In the case of $\mathfrak{g} = \mathfrak{sl}_n$, as described in [15], the cactus group $\mathrm{C}_n = \mathrm{C}_{\mathfrak{sl}_n}$ in Definition 2.5 can equivalently be expressed via the generators $c_{p,q}, 1 \leq p < q \leq n$ and relations:

- $c_{p,q}^2 = 1$,
- $c_{p,q}c_{m,r} = c_{m,r}c_{p,q}$ for any disjoint intervals $[p,q]$ and $[m,r]$,
- $c_{p,q}c_{m,r} = c_{p+q-r,p+q-m}c_{p,q}$ for any $[m,r]$ a subinterval of $[p,q]$.

We can then further restrict to a smaller generating set by noting that one can express $c_{p,q} = c_{1,q}c_{1,q+1-p}c_{1,q}$:

LEMMA 3.4. *The cactus group* C_n *is generated by the elements* $c_{1,q}$*, where* $2 \leq q \leq n$.

With the notation as above, consider the elements $a_{p,q,r} = c_{p,r}c_{p,q}c_{q+1,r}$ for $1 \leq p \leq q < r \leq n$, where we set $c_{p,p} = 1$. Then in particular $a_{p,p,p+1} = c_{p,p+1}$, $a_{p,q,r}^{-1} = a_{p,p+r-q-1,r}$, and the collection of elements $a_{p,q,r}$ generates C_n. They also determine a generating set for the pure cactus group:

PROPOSITION 3.5 ([**9**, Proposition 3.7]). *The conjugates of the elements*

$$b_{p,q,r,m} := a_{p,q,r}^{-1} a_{p+r-q,r,m}^{-1} a_{p,q,m}$$

for all $1 \leq p \leq q < r \leq n$ *generate the pure cactus group* PC_n.

EXAMPLE 3.6. For $\mathfrak{g} = \mathfrak{sl}_3$, the cactus group is given by

$$\mathrm{C}_3 = \left\langle c_{1,2}, c_{2,3}, c_{1,3} \mid c_{1,2}^2 = c_{2,3}^2 = c_{1,3}^2 = 1, c_{1,2}c_{1,3} = c_{1,3}c_{2,3} \right\rangle$$

Then, by Proposition 3.5 the element $b_{1,1,2,3} = c_{1,2}c_{2,3}c_{1,2}c_{1,3}$ and its conjugate $c_{2,3}b_{1,1,2,3}c_{2,3} = c_{2,3}c_{1,2}c_{2,3}c_{1,3} = b_{1,1,2,3}^{-1}$ generate $\mathrm{PC}_3 = \pi_1(\overline{\mathrm{M}}_{0,4}(\mathbb{R}), \mu) \cong \mathbb{Z}$. In the quotient of C_3 by PC_3, we have that $c_{1,3} \sim c_{1,2}c_{2,3}c_{1,2} \sim c_{2,3}c_{1,2}c_{2,3}$, so indeed $\mathrm{C}_3 \,/\, \mathrm{PC}_3 \cong S_3$.

4. Shift of argument algebras

We turn our attention next to a family of algebras associated to a semisimple Lie algebra $\mathfrak{g}$. We will work initially in $S(\mathfrak{g})$, the symmetric algebra of $\mathfrak{g}$. Recall that $S(\mathfrak{g})$ is $\mathbb{Z}_{\geq 0}$–graded, $S(\mathfrak{g}) = \bigoplus_{i \geq 0} S^i(\mathfrak{g})$, and has a Poisson bracket $\{\cdot, \cdot\}$ defined on degree one elements $x, y \in \mathfrak{g}$ as the Lie bracket, $\{x, y\} = [x, y]$, and extended via the Leibniz rule:

$$\{xy, z\} = x\{y, z\} + \{x, z\}y.$$

For any element $f \in S(\mathfrak{g}) = \mathbb{C}[\mathfrak{g}^*] \cong \mathbb{C}[\mathfrak{g}]$ and μ in $\mathfrak{g}^* \cong \mathfrak{g}$, one can consider the directional derivative of f in the μ direction, i.e. $\partial_\mu f \in S(\mathfrak{g})$, which for any $x \in \mathfrak{g}^* \cong \mathfrak{g}$ is given by

$$\partial_\mu f(x) = \frac{d}{dt} f(x + t\mu)|_{t=0}.$$

Then, to any $\mu \in \mathfrak{g}$ one can associate a subalgebra $A_\mu \subseteq S(\mathfrak{g})$, the **Mishchenko–Fomenko** or **shift of argument algebra**, whose elements are the μ–shifts of the Poisson center of $S(\mathfrak{g})$, $Z(S(\mathfrak{g})) = S(\mathfrak{g})^{\mathfrak{g}}$, the algebra of invariants under the adjoint action:

$$A_\mu = \left\langle \partial_\mu^i f \ : \ f \in S(\mathfrak{g})^{\mathfrak{g}}, i \geq 0 \right\rangle$$

Recall that $S(\mathfrak{g})^{\mathfrak{g}}$ is a free polynomial algebra with $\ell = \operatorname{rk} \mathfrak{g}$ algebraically independent generators. We can obtain a finite set of generators for A_μ by taking such a collection $\{f_1, f_2, \ldots, f_\ell\}$ of $S(\mathfrak{g})^{\mathfrak{g}}$ generators and all of their μ–translates. This produces $(\dim \mathfrak{g} + \operatorname{rk} \mathfrak{g})/2$ many generators of A_μ:

$$A_\mu = \left\langle \partial_\mu^{j_i} f_i, i = 1, \ldots, \ell, j_i = 0, \ldots, \deg f_i - 1 \right\rangle.$$

Here, $d_i = \deg f_i - 1$ are the exponents of $\mathfrak{g}$. Equivalently, one can take such a collection of basic invariants $\{f_i\}_{i=1}^{\operatorname{rk} \mathfrak{g}}$ and their shifts by $t\mu$ for all $t \in \mathbb{C}$.

EXAMPLE 4.1. For instance, take $\mathfrak{g} = \mathfrak{sl}_4$ and use the Killing form to identify $S(\mathfrak{g}) \cong \mathbb{C}[\mathfrak{g}]$. Then a collection of basic invariant polynomials is $f_1(x) = \operatorname{tr}(x^2), f_2(x) = \operatorname{tr}(x^3), f_3(x) = \operatorname{tr}(x^4)$, and for any $\mu \in \mathfrak{g}$ the shift of argument algebra A_μ has $(\operatorname{rk}\mathfrak{g} + \dim\mathfrak{g})/2 = (3+15)/2 = 9$ generators:

$$A_\mu = \left\langle \operatorname{tr}(\mu x), \operatorname{tr}(\mu^2 x), \operatorname{tr}(\mu^3 x), \operatorname{tr}(x^3), \operatorname{tr}(\mu x^2), \operatorname{tr}((\mu x)^2), \operatorname{tr}(x^2), \operatorname{tr}(\mu x^3), \operatorname{tr}(x^4)\right\rangle.$$

The shift of argument algebra A_μ, where $\mu \in \mathfrak{g}$, is Poisson-commutative. Moreover, for μ a regular element, A_μ satisfies the following properties originally shown in the Cartan setting by Shuvalov and Tarasov and later generalized by Panyushev–Yakimova to all of $\mathfrak{g}^{\mathrm{reg}}$.

THEOREM 4.2. *For any $\mu \in \mathfrak{g}^{reg}$, the shift of argument algebra A_μ is:*

- **[25, 31]** *a free polynomial algebra in the* $(\dim\mathfrak{g}+\operatorname{rk}\mathfrak{g})/2$ *generators* $\{\partial_\mu^{j_i} f_i\}$ *described above, it is of maximal possible transcendence degree, and is*
- **[25, 35]** *a maximal Poisson-commutative subalgebra of $S(\mathfrak{g})$, i.e. equal to its Poisson centralizer.*

The algebra A_μ also inherits a grading from that of $S(\mathfrak{g})$, given by:

$$A_\mu = \bigoplus_{i\geq 0} A_\mu^i, \quad \text{where } A_\mu^i = A_\mu \cap S^i(\mathfrak{g})$$

For any $\mu \in \mathfrak{h}^{\mathrm{reg}}$, the degree one and two components of A_μ are (see **[36]**):

$$A_\mu^1 = A_\mu \cap \mathfrak{g} = \mathfrak{h} \qquad A_\mu^2 = A_\mu \cap S^2(\mathfrak{g}) = Q_\mu \oplus S^2(\mathfrak{h})$$

where $Q_\mu \subseteq S^2(\mathfrak{g})$ is an ℓ–dimensional subspace given by

$$Q_\mu = \left\{ \sum_{\alpha\in\Delta_+} \frac{\alpha(h)}{\alpha(\mu)} e_\alpha f_\alpha \ : \ h \in \mathfrak{h} \right\}$$

REMARK 4.3. Note that for any $\mu \in \mathfrak{g}$, the algebra A_μ commutes with the centralizer $\mathfrak{g}_\mu$, so $A_\mu \subseteq S(\mathfrak{g})^{\mathfrak{g}_\mu}$ **[27**, Proposition 2**]**. A property that A_μ is expected to have, which is not yet known in general, is that it is maximal Poisson-commutative in $S(\mathfrak{g})^{\mathfrak{g}_\mu}$. Below we discuss this as well as related results, which indicate that shift of argument algebras indexed by Cartan elements have an operadic nature.

As in the setup for Lemma 2.3, given any $\mu_0 \in \mathfrak{h}$, take its derived centralizer $\mathfrak{g}'_{\mu_0} = [\mathfrak{g}_{\mu_0}, \mathfrak{g}_{\mu_0}]$. Since μ_0 is not necessarily regular, the algebra $A_{\mu_0} \subseteq S(\mathfrak{g})$ is not necessarily a maximal Poisson-commutative subalgebra of $S(\mathfrak{g})$, but we can 'pad it out' to make it maximal. For that purpose, take a regular Cartan element $\mu_1 \in \mathfrak{g}'_{\mu_0}$ and the corresponding shift of argument algebra $A_{\mu_1} \subseteq S(\mathfrak{g}_{\mu_0})$. Consider the subalgebra $A_{(\mu_0,\mu_1)} := \langle A_{\mu_0}, A_{\mu_1}\rangle \subseteq S(\mathfrak{g})$. Then:

PROPOSITION 4.4. *With the notation as in the previous paragraph:*

- **[35]** $A_{(\mu_0,\mu_1)}$ *is a maximal Poisson-commutative subalgera in $S(\mathfrak{g})$.*
- **[31]** $A_{(\mu_0,\mu_1)}$ *is a free polynomial algebra with* $\frac{1}{2}(\operatorname{rk}\mathfrak{g} + \dim\mathfrak{g})$ *many generators, given by a subset of the generators of A_{μ_0}, i.e. $\partial_{\mu_0}^{j_i} f_i$, taking the derivatives up to some lower order, and all the generators for A_{μ_1}.*
- **[14]** A_{μ_0} *is a maximal Poisson-commutative subalgebra of $S(\mathfrak{g})^{\mathfrak{g}_{\mu_0}}$ of maximal transcendence degree.*

The last result in Proposition 4.4 follows from the first two. Of interest, more generally, is to extend the family of maximal Poisson-commutative shift of argument algebras to a flat family parametrized by the de Concini–Procesi compactification of $\mathbb{P}(\mathfrak{h}^{\mathrm{reg}})$, which we explore in the following section.

5. Closure of the parameter space

5.1. Regular canonical sequences. Work of Shuvalov and Vinberg describes limits of shift of argument algebras and their quadratic components along curves given by so-called *regular sequences*, described below. It suffices to consider sequences of (finite) length at most $\ell = \operatorname{rk}\mathfrak{g}$.

DEFINITION 5.1. Consider a polynomial $\mu(t) = \mu_0 + \mu_1 t + \ldots + \mu_k t^k \in \mathfrak{h}[t]$, $k \leq \ell$, which we can also think of as a sequence $\vec{\mu} = (\mu_0, \mu_1, \ldots, \mu_k)$. Given such a sequence, let $\mathfrak{g}_i(\vec{\mu}) = \mathfrak{g}_i := \mathfrak{g}_{\mu_0} \cap \mathfrak{g}_{\mu_1} \cap \ldots \cap \mathfrak{g}_{\mu_i}$ denote the intersection of the centralizers of the first i terms, where $i = 0, \ldots, k$, and $\mathfrak{g}_{-1} = \mathfrak{g}$.

(1) $\vec{\mu}$ is called **regular** if $\mathfrak{g}_{\mu_0} \cap \mathfrak{g}_{\mu_1} \cap \ldots \cap \mathfrak{g}_{\mu_k} = \mathfrak{h}$. This is equivalent to requiring that $\mu(t) \in \mathfrak{h}^{\mathrm{reg}}$ for all sufficiently small $t \neq 0$.
(2) A regular sequence $\vec{\mu}$ is called **canonical** if for all $i = 0, \ldots, k$,
   - $\mu_i \in \mathfrak{g}_{i-1}^{\mathrm{ss}}$ (the semisimple part of $\mathfrak{g}_{i-1}$), and
   - the projection of μ_i to each simple component of $\mathfrak{g}_{i-1}$ is nonzero.
(3) Two canonical regular sequences $\vec{\mu} = (\mu_0, \mu_1, \ldots, \mu_k)$, $\vec{\nu} = (\nu_0, \nu_1, \ldots, \nu_m)$ are called **equivalent**, denoted $\vec{\mu} \sim \vec{\nu}$, if $k = m$ and for all $i = 0, 1, \ldots, k$,
   - $\mathfrak{g}_i(\vec{\mu}) = \mathfrak{g}_i(\vec{\nu})$, and
   - the projections of μ_i and ν_i onto the simple components of $\mathfrak{g}_{i-1}(\vec{\mu})$ are proportional.

Note that for any $\mu \in \mathfrak{h}^{\mathrm{reg}}$, the dimension $d_j := \dim A_\mu^j$ of the degree j component of the shift of argument algebra A_μ is the same. Therefore, for a regular curve $\mu(t)$ in the Cartan we can compute the limit $\lim_{t\to 0} A_{\mu(t)} = \bigoplus_{j\geq 0} A^j$ by taking the limit in each degree, inside the corresponding Grassmannian, namely $A^j := \lim_{t\to 0} A^j_{\mu(t)} \in \operatorname{Gr}(d_j, S^j(\mathfrak{g}))$ (see [**31**]). The result is a maximal Poisson-commutative subalgebra of $S(\mathfrak{g})$ with the same Hilbert series, and hence the same transcendence degree equal to $\frac{1}{2}(\operatorname{rk}\mathfrak{g} + \dim\mathfrak{g})$ as all of the shift of argument algebras corresponding to regular points. Building on a construction of Vinberg, Shuvalov gives an algorithmic recipe for computing this limit. Let $A(\vec{\mu}) \subseteq S(\mathfrak{g})$ denote the algebra generated by $S(\mathfrak{h})$ and the $\mathfrak{g}_{i-1}$–invariant elements of $S(\mathfrak{g}_{i-1})$ as well as all their μ_i–shifts for all $i = 0, \ldots, \ell$:

$$A(\vec{\mu}) = \left\langle \partial_{\mu_0}^{j_0} S(\mathfrak{g})^{\mathfrak{g}}, \partial_{\mu_1}^{j_1} S(\mathfrak{g}_0)^{\mathfrak{g}_0}, \ldots, \partial_{\mu_\ell}^{j_\ell} S(\mathfrak{g}_{\ell-1})^{\mathfrak{g}_{\ell-1}}, S(\mathfrak{h}) \;:\; j_0, j_1, \ldots, j_\ell \geq 0 \right\rangle$$

THEOREM 5.2. [**31**] *The limit shift of argument algebras* $\lim_{t\to 0} A_{\mu(t)}$ *for regular curves* $\mu(t)$ *are in bijection with the set of regular canonical sequences up to equivalence, since:*

i) *The limit shift of argument algebra corresponding to* $\mu(t)$ *as* $t \to 0$ *is a free commutative algebra and coincides with the algebra obtained by Shuvalov's construction,*

$$\lim_{t\to 0} A_{\mu(t)} = A(\vec{\mu}).$$

ii) *For any regular sequence* $\vec{\mu}$*, there is a canonical regular sequence* $\vec{\mu}^c$ *such that* $A(\vec{\mu}) = A(\vec{\mu}^c)$.

iii) *Given two canonical regular sequences* $\vec{\lambda}^c$ *and* $\vec{\nu}^c$, $A(\vec{\lambda}^c) = A(\vec{\nu}^c)$ *if and only if* $\vec{\lambda}^c \sim \vec{\nu}^c$.

We can then relate shift of argument algebras and their limits to the de Concini–Procesi model by, for instance, looking at their quadratic components.

THEOREM 5.3.

1) [**36**] *The limits of the subspaces* $Q_{\mu(t)}$ *along regular curves* $\mu(t)$ *as* $t \to 0$ *are in bijection with the equivalence classes of regular canonical sequences.*
2) [**1**] *The map* $q : \mathcal{M}_{\mathfrak{g}} \to \mathrm{Gr}(\ell, S^2(\mathfrak{g})), \mu \mapsto Q_\mu$ *extends uniquely to a closed embedding* $\overline{q} : \overline{\mathcal{M}}_{\mathfrak{g}} \to \mathrm{Gr}(\ell, S^2(\mathfrak{g}))$. *Therefore* $\overline{Q}_{\mathfrak{g}}$, *the closure of the family of quadratic spaces* $\{Q_\mu : \mu \in \mathcal{M}_{\mathfrak{g}}\}$, *is a smooth subvariety of* $\mathrm{Gr}(\ell, S^2(\mathfrak{g}))$ *isomorphic to the de Concini–Procesi compactification* $\overline{\mathcal{M}}_{\mathfrak{g}}$.

REMARK 5.4. Shift of argument algebras indexed by regular Cartan elements (outside of a countable union of proper closed subvarieties of $\mathfrak{g}$) are determined by their quadratic components [**28**]. Theorem 5.2 and Theorem 5.3, part 1), imply that this property also extends when taking the corresponding limits. Together with Theorem 5.3, part 2), we get that the family of shift of argument algebras and their limits is indexed by the minimal de Concini–Procesi model $\overline{\mathcal{M}}_{\mathfrak{g}}$.

We consider next in a broader context the closure of a family of algebras that is parametrized by a variety.

DEFINITION 5.5. [See for instance [**14**, Definition 6.2], [**16**, Section 6.2]] Consider a $\mathbb{Z}_{\geq 0}$–graded commutative algebra $G = \bigoplus_{i \geq 0} G^i$ where each graded component G^i is finite-dimensional. We say a family of subalgebras of G is **parametrized by a variety** X if for any $\mu \in X$ there is a subalgebra A_μ of G such that for all $i \geq 0$, $a_i = \dim(A_\mu \cap G^i)$ is independent of μ and the map

$$g_i : X \to \prod_{j=0}^{i} \mathrm{Gr}(a_j, G^j), \quad \mu \mapsto (A_\mu \cap G^0, A_\mu \cap G^1, \ldots, A_\mu \cap G^i)$$

is a morphism of varieties. We can then take the **closure of this family** by first taking for each $i \in \mathbb{Z}_{\geq 0}$ the closure of the image of g_i,

$$Y_i := \overline{\mathrm{Im}\, g_i} \subseteq \prod_{j=0}^{i} \mathrm{Gr}(a_j, G^j).$$

We get a sequence of projections,

$$Y_0 \xleftarrow{p_1} Y_1 \xleftarrow{p_2} Y_2 \xleftarrow{p_3} \ldots$$

The *closure of the family of algebras* is then obtained as the inverse limit $Y = \varprojlim Y_i$, which is well-defined as a pro-algebraic scheme and is naturally a parameter space for a larger family of $\mathbb{Z}_{\geq 0}$–graded commutative algebras (since being a subalgebra and being commutative are both Zariski-closed conditions on the parameters). A point $\vec{y} \in Y$ is a sequence $\vec{y} = \{y_i \in Y_i\}_{i \geq 0}$ where $y_i = (y_i^0, y_i^1, \ldots, y_i^i) \in \prod_{j=0}^{i} \mathrm{Gr}(a_j, G^j)$ and such that $p_i(y_i) = y_{i-1}$ for all i. The point $\vec{y}$ corresponds to an algebra $A(\vec{y}) = \bigoplus_{i \geq 0} y_i^i = \bigoplus_{i \geq 0} A^i(\vec{y})$.

5.2. An isomorphism of varieties. Let $\overline{\mathfrak{A}}_{\mathfrak{g}}$ denote the closure of the family of shift of argument algebras parametrized by $\mathcal{M}_{\mathfrak{g}} = \mathbb{P}(\mathfrak{h}^{\text{reg}})$, as described in Definition 5.5.

THEOREM 5.6. *$\overline{\mathfrak{A}}_{\mathfrak{g}}$ is a smooth variety isomorphic to $\overline{\mathcal{M}_{\mathfrak{g}}}$. In other words, the map that assigns to each $\mu \in \mathcal{M}_{\mathfrak{g}}$ a shift of argument algebra A_μ can be extended over $\overline{\mathcal{M}_{\mathfrak{g}}}$, parametrizing a family of maximal Poisson-commutative subalgebras of $S(\mathfrak{g})$.*

This result essentially follows from Shuvalov's work discussed above. We recall the proof in [**14**].

PROOF. Consider the closure $\overline{\mathfrak{A}}_{\mathfrak{g}} := Y = \lim\limits_{\leftarrow} Y_i$, as in Definition 5.5, of the family of shift of argument algebras parametrized by $\mathcal{M}_{\mathfrak{g}}$. Here, each Y_i is a projective variety and by Theorem 5.3, we have an isomorphism $Y_2 \cong \overline{Q}_{\mathfrak{g}} \cong \overline{\mathcal{M}_{\mathfrak{g}}}$. Moreover, by Theorem 5.2, the natural map $Y \to Y_2$ is a bijection on $\mathbb{C}$-points. Therefore, the morphisms $Y_{i+1} \to Y_i$ are also bijections for all $i \geq 2$. Since Y_2 is smooth, we get isomorphisms $Y_2 \cong Y_i$ for all $i \geq 3$. Therefore, we also have an isomorphism $Y \cong Y_2 \cong \overline{\mathcal{M}_{\mathfrak{g}}}$. □

5.3. Lifting to $\mathcal{U}(\mathfrak{g})$ and the operadic nature of shift of argument algebras. In [**27**] and [**28**], Rybnikov shows that shift of argument algebras in $S(\mathfrak{g})$ can be lifted uniquely to $\mathcal{U}(\mathfrak{g})$. This was originally shown by Tarasov in the case $\mathfrak{g} = \mathfrak{sl}_n$ through the symmetrization map $S(\mathfrak{g}) \to \mathcal{U}(\mathfrak{g})$ (see [**33, 34**]).

THEOREM 5.7 ([**27, 28**]). *For any semisimple Lie algebra $\mathfrak{g}$, the family of shift of argument algebras $A_\mu \subseteq S(\mathfrak{g})$ for $\mu \in \mathbb{P}(\mathfrak{h})$ can be lifted uniquely to a family of commutative subalgebras $\mathcal{A}_\mu \subseteq \mathcal{U}(\mathfrak{g})$, known as the quantum shift of argument algebras, such that* $\operatorname{gr} \mathcal{A}_\mu = A_\mu$.

EXAMPLE 5.8. (See for instance [**27**, Lemma 4].) In the case of $\mathfrak{g} = \mathfrak{sl}_n$, consider the regular curve $\mu(t) = \mu_1 + \mu_2 t + \ldots + \mu_n t^{n-1}$ where $\mu_i = E_{1,1} + E_{2,2} + \ldots + E_{n-i,n-i} - (n-i)E_{n,n}$. Then the limit subalgebra $\lim_{t \to 0} \mathcal{A}_{\mu(t)}$ is the *Gelfand–Tsetlin subalgebra* in $\mathcal{U}(\mathfrak{sl}_n)$, a maximal commutative subalgebra generated by the centers $Z(\mathcal{U}(\mathfrak{sl}_n)), Z(\mathcal{U}(\mathfrak{sl}_{n-1})), \ldots, Z(\mathcal{U}(\mathfrak{sl}_2))$ for a corresponding sequence $\mathfrak{sl}_n \supseteq \mathfrak{sl}_{n-1} \supseteq \ldots \supseteq \mathfrak{sl}_2$.

One can define, analogously to Definition 5.5, the closure of a family of filtered algebras parametrized by a variety. With this definition, we can consider the closure of the family of quantum shift of argument algebras.

THEOREM 5.9 ([**14**, Theorem 10.7]). *The closure of the family of quantum shift of argument algebras $\mathcal{A}_\mu$ parametrized by $\mu \in \mathcal{M}_{\mathfrak{g}}$ is a smooth projective variety isomorphic to the de Concini–Procesi model $\overline{\mathcal{M}_{\mathfrak{g}}}$.*

The proof of this theorem involves considering a collection of charts on $\overline{\mathcal{M}_{\mathfrak{g}}}$ and constructing a set of generators for the corresponding shift of argument algebras on a given chart.

This family of quantum shift of argument algebras also has a particular operadic nature, which will be of importance in Section 6 when considering how it acts on a given highest-weight irreducible $\mathfrak{g}$-representation. Namely, with the notation as in Lemma 2.3 and using Proposition 4.4, we get that the following holds:

PROPOSITION 5.10 ([**14**, Corollary 10.12]). *For any $\mu_0 \neq 0$ in $\mathfrak{h}$ and $\mu_1 \in \overline{\mathcal{M}}_{\mathfrak{g}'_{\mu_0}}$, considering $\mu = (\mu_0, \mu_1)$ as a point in $\overline{\mathcal{M}_{\mathfrak{g}}}$, the algebra $\mathcal{A}_\mu$ is generated by $\mathcal{A}_{\mu_0}, \mathcal{A}_{\mu_1}$, and has the following structure:*

$$\mathcal{A}_\mu = \mathcal{A}_{\mu_0} \otimes_{Z(U(\mathfrak{g}_1))} \mathcal{A}_{\mu_1}.$$

6. A monodromy action, the cactus group, and $\mathfrak{g}$-crystals

6.1. A cactus group action on crystals. Some of the central objects of combinatorial representation theory are $\mathfrak{g}$-crystals, which reflect the structure of representations of the quantum group of $\mathfrak{g}$. A crystal corresponding to such a representation V is a set B_V, which can be thought of as a collection of elements forming a 'nice' basis for V, together with maps $e_i, f_i : B_V \to B_V \cup \{0\}$, $i \in I$, called the Kashiwara operators and obtained from the Chevalley generators of $\mathfrak{g}$, as well as a weight map $\mathrm{wt} : B_V \to \Lambda$. Here, I is the Dynkin diagram of $\mathfrak{g}$ as before and Λ is the weight lattice. There is a set-theoretic action of the cactus group $C_{\mathfrak{g}}$ on any $\mathfrak{g}$-crystal $B = B_V$ defined via generalized Schützenberger involutions as follows. Pictured as a colored directed graph, $B = B_V = \bigsqcup_\lambda B_\lambda$ is the disjoint union of a collection of connected components, corresponding to the decomposition of V into highest-weight irreducible representations. Then the involution $\xi = \xi_I$ acts as ξ_λ on each B_λ, where ξ_λ is uniquely determined by the following properties, for all $b \in B_\lambda$, $i \in I$:

1) $\mathrm{wt}(\xi_\lambda(b)) = w_0 \,\mathrm{wt}(b)$,
2) $\xi(e_i(b)) = f_{\theta_I(i)}(\xi(b))$,
3) $\xi(f_i(b)) = e_{\theta_I(i)}(\xi(b))$.

The map $\theta_I : I \to I$ is the Dynkin diagram automorphism as in Definition 2.5. Similarly, for any connected $J \subseteq I$, one can define ξ_J by first decomposing the restriction $B|_J$ (i.e. erasing all arrows in B labelled by $I \setminus J$) into irreducible components. As seen by their defining relations, the ξ_J's do not preserve the crystal structure, but rather can be seen as automorphisms of the forgetful functor from $\mathfrak{g}$-crystals to sets. To summarize:

THEOREM 6.1. *There is an action of $C_{\mathfrak{g}}$ on any $\mathfrak{g}$-crystal B as a set, given by the assignment $c_J \mapsto \xi_J$, for any connected subdiagram $J \subseteq I$.*

In the next section, we relate this cactus action to shift of argument algebras.

6.2. Covering spaces of $\overline{\mathcal{M}}_{\mathfrak{g}}(\mathbb{R})$ and crystal structures. For a fixed irreducible highest weight $\mathfrak{g}$-representation $V(\lambda)$, the action of the shift of argument algebras on $V(\lambda)$ satisfies several nice properties:

PROPOSITION 6.2 ([**14**]). *For any $\mu \in \overline{\mathcal{M}}_{\mathfrak{g}}(\mathbb{R})$, the shift of argument algebra $\mathcal{A}_\mu$ has simple spectrum on $V(\lambda)$. If we take $E(\lambda) = \{(\mu, \ell) \ : \ \mu \in \overline{\mathcal{M}}_{\mathfrak{g}}(\mathbb{R}), \ell \in \mathbb{P}(V(\lambda))$ is an eigenline of $\mathcal{A}_\mu\}$, then $E(\lambda) \xrightarrow{\pi_\lambda} \overline{\mathcal{M}}_{\mathfrak{g}}(\mathbb{R})$ is an unramified covering.*

This proposition can be proved by showing that when working with a real parameter $\mu \in \overline{\mathcal{M}}_{\mathfrak{g}}(\mathbb{R})$ (see [**14**, Section 11] and [**30**]):

- the $\mathfrak{g}$-representation $V(\lambda)$ is a cyclic $\mathcal{A}_\mu$-module, i.e. there is an element $v \in V(\lambda)$ such that $V(\lambda) = \mathcal{A}_\mu \cdot v$, and
- $\mathcal{A}_\mu$ acts by Hermitian operators on $V(\lambda)$, for the Hermitian form preserved by the compact form of $\mathfrak{g}$.

We will next describe a crystal structure on a fiber of this covering. Take any μ in the open dominant Weyl chamber, as described in Section 2.3, and consider the fiber

$$E(\lambda,\mu) = E_{\mathcal{A}_\mu}(V(\lambda)) := \pi_\lambda^{-1}(\mu).$$

We start by defining the weight map wt : $E(\lambda,\mu) \to \Lambda$. Since $\mathfrak{h} \subseteq \mathcal{A}_\mu$, the spectrum of each shift of argument algebra respects the weight decomposition of $V(\lambda)$, i.e. every eigenline ℓ for $\mathcal{A}_\mu \curvearrowright V(\lambda)$ lives in some weight space $V(\lambda)_\xi$, and we set $\mathrm{wt}(\ell) := \xi$.

We will next describe the construction of the raising Kashiwara crystal operator $e_i : E(\lambda,\mu) \to E(\lambda,\mu) \sqcup \{0\}$ for a given $i \in I$. The definition of f_i is similar. We begin by giving the construction on the $\{i\}$-facet of $\overline{P}_+$ (see the notation in Section 2.3), corresponding to the vanishing of the i-th simple root, and then use parallel transport to extend it to the fiber over μ. Namely, take a point

$$\mu^i \in \overline{\mathcal{M}}_{\mathfrak{g}}(\mathbb{R})^{\{i\}} := \overline{P}_+ \cap p^{-1}(\{\nu \in \mathfrak{h}(\mathbb{R}) \ : \ \alpha_i(\nu) = 0, \alpha_j(\nu) > 0 \text{ for } j \neq i\}).$$

We can write $\mu^i = (\mu_0^i, \mu_1^i)$ as in Proposition 5.10, where

$$\mu_0^i \in \{\nu \in \mathfrak{h}(\mathbb{R}) \ : \ \alpha_i(\nu) = 0, \alpha_j(\nu) > 0 \text{ for } j \neq i\},$$

so $\mathfrak{g}'_{\mu_0^i} \cong (\mathfrak{sl}_2)_i$, the $\mathfrak{sl}_2$ Lie subalgebra of $\mathfrak{g}$ corresponding to the simple root α_i, and $\mu_1^i \in \overline{\mathcal{M}}_{(\mathfrak{sl}_2)_i}(\mathbb{R}) = \{\mathrm{pt}\}$.

As a further consequence of Proposition 5.10, the action of $\mathcal{A}_{\mu^i}$ respects the decomposition of $V(\lambda)$ under $(\mathfrak{sl}_2)_i$:

$$V(\lambda) = \bigoplus_\nu \mathrm{Hom}_{(\mathfrak{sl}_2)_i}(V(\nu), V(\lambda)) \otimes V(\nu)$$

where ν ranges over dominant $(\mathfrak{sl}_2)_i$-weights. So, on the level of fibers for the action on the left side of the equation, $\mathcal{U}(\mathfrak{g}) \supseteq \mathcal{A}_{\mu^i} \curvearrowright V(\lambda)$, and the action on the right side $\mathcal{U}(\mathfrak{g})^{(\mathfrak{sl}_2)_i} \supseteq \mathcal{A}_{\mu_0^i} \curvearrowright \mathrm{Hom}_{(\mathfrak{sl}_2)_i}(V(\nu), V(\lambda))$ and $\mathcal{U}((\mathfrak{sl}_2)_i) \supseteq \mathcal{A}_{\mu_1^i} \curvearrowright V(\nu)$, we get:

$$(*) \qquad E_{\mathcal{A}_{\mu^i}}(V(\lambda)) = \bigsqcup_\nu E_{\mathcal{A}_{\mu_0^i}}(\mathrm{Hom}_{(\mathfrak{sl}_2)_i}(V(\nu), V(\lambda))) \times E_{\mathcal{A}_{\mu_1^i}}(V(\nu))$$

From the earlier discussion, we can deduce that the set

$$E(\mu_1^i, \nu) = E_{\mathcal{A}_{\mu_1^i}}(V(\nu))$$

is just the set of one-dimensional $(\mathfrak{sl}_2)_i$-weightspaces of $V(\nu)$. So, for a pair (ℓ_1, ℓ_2) as in Equation (*), the $(\mathfrak{sl}_2)_i$ crystal operator e_{α_i} acts as $e_{\alpha_i}(\ell_1, \ell_2) = (\ell_1, e_{\alpha_i}(\ell_2))$, where ℓ_2 is sent to another eigenline or to zero depending on its weight. We can then parallel transport the fiber $E(\lambda,\mu)$ to $E(\lambda,\mu^i)$ using the path γ_{μ,μ^i} discussed in Section 2.3, apply e_{α_i}, and then transport it back to define the crystal operator $e_i : E(\lambda,\mu) \to E(\lambda,\mu) \sqcup \{0\}$,

$$e_i = \gamma_{\mu,\mu^i}^{-1} \circ e_{\alpha_i} \circ \gamma_{\mu,\mu^i}.$$

Some properties one needs to check in order to confirm that this indeed defines a crystal structure on $E(\lambda,\mu)$ are that it is is compatible with taking tensor product as well as with restricting to Levi subalgebras (see [**14**, Section 12]).

THEOREM 6.3 ([**14**]). *With the above described crystal structure on* $E(\lambda, \mu)$, *there is a crystal isomorphism* $E(\lambda, \mu) \cong B(\lambda)$. *Under this crystal isomorphism, the monodromy action* $\pi_1^W(\overline{\mathcal{M}}_{\mathfrak{g}}(\mathbb{R}), \mu) \curvearrowright E(\lambda, \mu)$ *agrees with the cactus action* $C_{\mathfrak{g}} \curvearrowright B(\lambda)$ *described in Section* 6.1.

The connection between shift of argument algebras and their generalizations on one hand, and de Concini–Procesi compactifications and their variations on the other, extends further in the context of the representation theory of the corresponding Lie algebra $\mathfrak{g}$. The reader is encouraged to look for instance at [**37**], [**32**], [**14**], and [**22**] for some further information.

References

[1] Leonardo Aguirre, Giovanni Felder, and Alexander P. Veselov, *Gaudin subalgebras and wonderful models*, Selecta Math. (N.S.) **22** (2016), no. 3, 1057–1071, DOI 10.1007/s00029-015-0213-y. MR3518545

[2] Bojko Bakalov and Alexander Kirillov Jr., *Lectures on tensor categories and modular functors*, University Lecture Series, vol. 21, American Mathematical Society, Providence, RI, 2001, DOI 10.1090/ulect/021. MR1797619

[3] A. N. Kirillov and A. D. Berenstein, *Groups generated by involutions, Gel′fand-Tsetlin patterns, and combinatorics of Young tableaux*, Algebra i Analiz **7** (1995), no. 1, 92–152; English transl., St. Petersburg Math. J. **7** (1996), no. 1, 77–127. MR1334154

[4] M. Davis, T. Januszkiewicz, and R. Scott, *Fundamental groups of blow-ups*, Adv. Math. **177** (2003), no. 1, 115–179, DOI 10.1016/S0001-8708(03)00075-6. MR1985196

[5] C. De Concini and C. Procesi, *Wonderful models of subspace arrangements*, Selecta Math. (N.S.) **1** (1995), no. 3, 459–494, DOI 10.1007/BF01589496. MR1366622

[6] C. De Concini and C. Procesi, *Hyperplane arrangements and holonomy equations*, Selecta Math. (N.S.) **1** (1995), no. 3, 495–535, DOI 10.1007/BF01589497. MR1366623

[7] P. Deligne and D. Mumford, *The irreducibility of the space of curves of given genus*, Inst. Hautes Études Sci. Publ. Math. **36** (1969), 75–109. MR262240

[8] Satyan L. Devadoss, *Tessellations of moduli spaces and the mosaic operad*, Homotopy invariant algebraic structures (Baltimore, MD, 1998), Contemp. Math., vol. 239, Amer. Math. Soc., Providence, RI, 1999, pp. 91–114, DOI 10.1090/conm/239/03599. MR1718078

[9] Pavel Etingof, André Henriques, Joel Kamnitzer, and Eric M. Rains, *The cohomology ring of the real locus of the moduli space of stable curves of genus 0 with marked points*, Ann. of Math. (2) **171** (2010), no. 2, 731–777, DOI 10.4007/annals.2010.171.731. MR2630055

[10] A. S. Miščenko and A. T. Fomenko, *Integrability of Euler's equations on semisimple Lie algebras* (Russian), Trudy Sem. Vektor. Tenzor. Anal. **19** (1979), 3–94. MR549008

[11] Giovanni Gaiffi, *Real structures of models of arrangements*, Int. Math. Res. Not. **64** (2004), 3439–3467, DOI 10.1155/S1073792804141172. MR2101279

[12] Victor Ginzburg and Mikhail Kapranov, *Koszul duality for operads*, Duke Math. J. **76** (1994), no. 1, 203–272, DOI 10.1215/S0012-7094-94-07608-4. MR1301191

[13] Iva Halacheva, *Alexander Type Invariants of Tangles, Skew Howe Duality for Crystals and The Cactus Group*, ProQuest LLC, Ann Arbor, MI, 2016. Thesis (Ph.D.)–University of Toronto (Canada). MR3653607

[14] Iva Halacheva, Joel Kamnitzer, Leonid Rybnikov, and Alex Weekes, *Crystals and monodromy of Bethe vectors*, Duke Math. J. **169** (2020), no. 12, 2337–2419, DOI 10.1215/00127094-2020-0003. MR4139044

[15] André Henriques and Joel Kamnitzer, *Crystals and coboundary categories*, Duke Math. J. **132** (2006), no. 2, 191–216, DOI 10.1215/S0012-7094-06-13221-0. MR2219257

[16] A. Ilin and L. Rybnikov, *On classical limits of Bethe subalgebras in Yangians*, Transform. Groups **26** (2021), no. 2, 537–564, DOI 10.1007/s00031-021-09648-x. MR4404439

[17] Mikhail M. Kapranov, *The permutoassociahedron, Mac Lane's coherence theorem and asymptotic zones for the KZ equation*, J. Pure Appl. Algebra **85** (1993), no. 2, 119–142, DOI 10.1016/0022-4049(93)90049-Y. MR1207505

[18] Sean Keel, *Intersection theory of moduli space of stable n-pointed curves of genus zero*, Trans. Amer. Math. Soc. **330** (1992), no. 2, 545–574, DOI 10.2307/2153922. MR1034665

[19] A, Khoroshkin and T. Willwacher, *Real moduli space of stable rational curves revisited*, Preprint, arXiv:1905.04499.

[20] Finn F. Knudsen, *The projectivity of the moduli space of stable curves. II. The stacks $M_{g,n}$*, Math. Scand. **52** (1983), no. 2, 161–199, DOI 10.7146/math.scand.a-12001. MR702953

[21] Bertram Kostant, *Lie group representations on polynomial rings*, Amer. J. Math. **85** (1963), 327–404, DOI 10.2307/2373130. MR158024

[22] V. Krylov, I. Mashanova-Golikova, and L. Rybnikov, *Bethe subalgebras in Yangians and Kirillov–Reshetikin crystals*, Preprint, arXiv:2212.11995.

[23] Ivan Losev, *Cacti and cells*, J. Eur. Math. Soc. (JEMS) **21** (2019), no. 6, 1729–1750, DOI 10.4171/JEMS/871. MR3945740

[24] Martin Markl, Steve Shnider, and Jim Stasheff, *Operads in algebra, topology and physics*, Mathematical Surveys and Monographs, vol. 96, American Mathematical Society, Providence, RI, 2002, DOI 10.1090/surv/096. MR1898414

[25] Dmitri I. Panyushev and Oksana S. Yakimova, *The argument shift method and maximal commutative subalgebras of Poisson algebras*, Math. Res. Lett. **15** (2008), no. 2, 239–249, DOI 10.4310/MRL.2008.v15.n2.a3. MR2385637

[26] Eric M. Rains, *The homology of real subspace arrangements*, J. Topol. **3** (2010), no. 4, 786–818, DOI 10.1112/jtopol/jtq027. MR2746338

[27] L. G. Rybnikov, *The shift of invariants method and the Gaudin model* (Russian, with Russian summary), Funktsional. Anal. i Prilozhen. **40** (2006), no. 3, 30–43, 96, DOI 10.1007/s10688-006-0030-3; English transl., Funct. Anal. Appl. **40** (2006), no. 3, 188–199. MR2265683

[28] L. G. Rybnikov, *Centralizers of some quadratic elements in Poisson-Lie algebras and a method for the translation of invariants* (Russian), Uspekhi Mat. Nauk **60** (2005), no. 2(362), 173–174, DOI 10.1070/RM2005v060n02ABEH000840; English transl., Russian Math. Surveys **60** (2005), no. 2, 367–369. MR2152960

[29] Leonid Rybnikov, *Cactus group and monodromy of Bethe vectors*, Int. Math. Res. Not. IMRN **1** (2018), 202–235, DOI 10.1093/imrn/rnw259. MR3801430

[30] Leonid Rybnikov, *A proof of the Gaudin Bethe ansatz conjecture*, Int. Math. Res. Not. IMRN **22** (2020), 8766–8785, DOI 10.1093/imrn/rny245. MR4216703

[31] V. V. Shuvalov, *On the limits of Mishchenko-Fomenko subalgebras in Poisson algebras of semisimple Lie algebras* (Russian, with Russian summary), Funktsional. Anal. i Prilozhen. **36** (2002), no. 4, 55–64, DOI 10.1023/A:1021713927119; English transl., Funct. Anal. Appl. **36** (2002), no. 4, 298–305. MR1958995

[32] David E. Speyer, *Schubert problems with respect to osculating flags of stable rational curves*, Algebr. Geom. **1** (2014), no. 1, 14–45, DOI 10.14231/AG-2014-002. MR3234112

[33] A. A. Tarasov, *On some commutative subalgebras in the universal enveloping algebra of the Lie algebra* $\mathfrak{gl}(n, \mathbf{C})$ (Russian, with Russian summary), Mat. Sb. **191** (2000), no. 9, 115–122, DOI 10.1070/SM2000v191n09ABEH000509; English transl., Sb. Math. **191** (2000), no. 9-10, 1375–1382. MR1805600

[34] A. A. Tarasov, *On the uniqueness of the lifting of maximal commutative subalgebras of the Poisson-Lie algebra to the enveloping algebra* (Russian, with Russian summary), Mat. Sb. **194** (2003), no. 7, 155–160, DOI 10.1070/SM2003v194n07ABEH000757; English transl., Sb. Math. **194** (2003), no. 7-8, 1105–1111. MR2020383

[35] A. A. Tarasov, *The maximality of some commutative subalgebras in Poisson algebras of semisimple Lie algebras* (Russian), Uspekhi Mat. Nauk **57** (2002), no. 5(347), 165–166, DOI 10.1070/RM2002v057n05ABEH000567; English transl., Russian Math. Surveys **57** (2002), no. 5, 1013–1014. MR1992097

[36] È. B. Vinberg, *Some commutative subalgebras of a universal enveloping algebra* (Russian), Izv. Akad. Nauk SSSR Ser. Mat. **54** (1990), no. 1, 3–25, 221, DOI 10.1070/IM1991v036n01ABEH001925; English transl., Math. USSR-Izv. **36** (1991), no. 1, 1–22. MR1044045

[37] Noah White, *The monodromy of real Bethe vectors for the Gaudin model*, J. Comb. Algebra **2** (2018), no. 3, 259–300, DOI 10.4171/JCA/2-3-3. MR3845719

Northeastern University, Department of Mathematics, 360 Huntington Ave, Boston, Massachusetts 02115

Email address: i.halacheva@northeastern.edu

Contemporary Mathematics
Volume **790**, 2023
https://doi.org/10.1090/conm/790/15860

Projection spaces and twisted Lie algebras

Ben Knudsen

ABSTRACT. A projection space is a collection of spaces interrelated by the combinatorics of projection onto tensor factors in a symmetric monoidal background category. Examples include classical configuration spaces, orbit configuration spaces, the graphical configuration spaces of Eastwood–Huggett, the simplicial configuration spaces of Cooper–de Silva–Sazdanovic, the generalized configuration spaces of Petersen, and Stiefel manifolds. We show that, under natural assumptions on the background category, the homology of a projection space is calculated by the Chevalley–Eilenberg complex of a certain generalized Lie algebra. We identify conditions on this Lie algebra implying representation stability in the classical setting of finite sets and injections.

1. Introduction

Experience has proven the value of studying configuration spaces in families. When organized correctly, these spaces exhibit emergent algebraic structure that imposes strong constraints on their topological invariants. One important means of organization involves the background space in an essential way; one speaks of operads and their algebras and modules [**22, 28**], or of adding a particle near the boundary of a manifold [**29**] or onto an edge of a graph [**2**]. Here, we pursue an orthogonal organizing principle, namely that of the underlying combinatorics.

1.1. Context and motivation. As a motivating example, consider the ordinary configuration space $\mathrm{Conf}_I(X)$, defined as the space of injections from the finite set I into the topological space X. The composite of injections being an injection, the collection of all such spaces forms a presheaf on the category $\mathcal{FI}$ of finite sets and injections. In many examples of interest, this combinatorial structure forces the rational (co)homology of configuration spaces to exhibit *representation stability* [**6**].

There are now a number of machines devoted to the study of stability phenomena in various contexts [**19, 35**]. Unfortunately, most of these machines are adapted to the study of automorphism groups, rather than configuration spaces. The goal of this paper is to develop a framework better adapted to examples such as those listed in the following table.

We proceed from the observation that the projection $\mathrm{Conf}_J(X) \to \mathrm{Conf}_I(X)$ induced by the injection $f : I \to J$ is subsumed by the map that splits an I-indexed configuration into a configuration indexed by the image of f and one indexed by

Combinatorics	Configuration space
Sets	Ordinary
G-sets	Orbit
Graphs	Graphical [**10**]
Simplicial complexes	Simplicial [**7**]
Collision structures	Generalized [**32**]
Vector spaces	Stiefel

its complement. The collection of all such splitting maps can be regarded as a kind of cocommutative comultiplication, which, according to the philosophy of Koszul duality, is governed by its Lie algebra of (derived) primitives.

The resulting interpretation of the homology of ordinary configuration spaces as Lie algebra homology has proven quite fruitful [**4, 9, 15, 23–25**]. This success motivates us to generalize the relationship, showing that the homology of something sufficiently like a configuration space is governed by something like a Lie algebra.

1.2. Framework and results. Our first step is to generalize the relationship between bijections and injections among finite sets in the relationship between a symmetric monoidal category $\mathcal{C}$ and its projection category $\mathrm{Pr}(\mathcal{C})$. Roughly, a morphism in $\mathrm{Pr}(\mathcal{C})$ is (opposite to) a projection onto a tensor factor in $\mathcal{C}$ (see Section 2.1 for details). We then make the following definition, which encompasses all of the spaces listed in the table above (see Section 6.1).

Definition 1.1. A *projection space* over $\mathcal{C}$ is a functor $X : \mathrm{Pr}(\mathcal{C})^{\mathrm{op}} \to \mathcal{T}\mathrm{op}$. We say that X is *reduced* if its value on the monoidal unit is a singleton.

The bulk of the paper is spent in constructing the following composite functor, which associates to each projection space X a $\mathcal{C}$-*twisted Lie algebra* $L(X)$ of *rational primitives*:

$$\begin{array}{l} L : \mathrm{Fun}(\mathrm{Pr}(\mathcal{C})^{\mathrm{op}}, \mathcal{T}\mathrm{op}) \xrightarrow{\cong} \mathrm{Fun}^{\mathrm{oplax}}(\mathcal{C}^{\mathrm{op}}, \mathcal{T}\mathrm{op}) \\ \qquad \downarrow A_* \\ \mathrm{Fun}^{\mathrm{oplax}}(\mathcal{C}^{\mathrm{op}}, \mathcal{C}\mathrm{h}_{\mathbb{Q}}) \xrightarrow{\cong} \mathrm{Coalg}_{\mathrm{Com}}(\mathrm{Fun}(\mathcal{C}^{\mathrm{op}}, \mathcal{C}\mathrm{h}_{\mathbb{Q}})) \\ \qquad \downarrow \mathcal{Q} \\ \mathrm{Alg}_{\mathrm{Lie}}(\mathrm{Fun}(\mathcal{C}^{\mathrm{op}}, \mathcal{C}\mathrm{h}_{\mathbb{Q}})). \end{array}$$

Briefly, the first functor witnesses a universal property of the projection category (Corollary 4.3); the second is the functor of Sullivan chains, a suitable rational replacement for singular chains (Corollary A.5); the third relies on a variant of Day convolution (Corollary 4.12); and the fourth is Quillen's Koszul duality functor [**33**]. The main result is as follows—see Section 5 for undefined terms.

Theorem 1.2. *Let X be a reduced projection space over the combinatorial symmetric monoidal category $\mathcal{C}$. There is a canonical isomorphism of $\mathcal{C}$-twisted cocommutative coalgebras*

$$H_*(X; \mathbb{Q}) \cong H_*^{\mathrm{Lie}}(L(X)).$$

As a consequence, the homology of generalized configuration spaces is calculated by the Chevalley–Eilenberg complex of a certain Lie algebra. This general connection to Lie algebras may at first seem surprising. Indeed, for ordinary configuration spaces, it is natural to view the connection as arising by combining the Goresky–MacPherson formula, expressing the homology of a stratified space in terms of poset homology, with the identification of the Lie operad with the homology of the partition posets [**32**]. In the more general setting, the resulting poset homology has no obvious relation to the Lie operad, yet the connection persists.

As foreshadowed above, we imagine Theorem 1.2 as a source and organizing principle for stability phenomena. As a proof of concept, we show that our framework encompasses representation stability. Specifically, stability is tied to the eventual high connectivity of the Lie algebra of rational primitives.

THEOREM 1.3. *Let X be a reduced $\mathcal{FI}^{\mathrm{op}}$-space taking values in path connected spaces. Suppose that $L(X)$ satisfies the following conditions.*

(1) *$H_i(L(X)_k)$ is finite dimensional for every $i \geq 0$ and $k \geq 0$.*
(2) *$H_0(L(X)_k) = 0$ for $k > 1$.*
(3) *$H_i(L(X)_k) = 0$ for i fixed and k sufficiently large.*

Then $H^*(X; \mathbb{Q})$ is representation stable.

We deduce this theorem from a more general result encompassing projection spaces with path disconnected values—see Theorem 6.14.

1.3. Future directions. Our work leaves a rather large number of natural questions unanswered (but see [**20**] and [**21**] for closely related work). The following three extensions at least are likely within easy reach.

(1) *Integer coefficients.* Our program relies crucially on a strictly symmetric replacement for the oplax monoidal functor of singular chains. In characteristic zero, such a replacement is within easy reach (see Appendix A). With a bit less laziness, one should be able to work integrally, as in [**32**], after minor modifications of the techniques of [**36**]—see also Remark A.6.

(2) *Classical primitives.* Let M be a (for simplicity) orientable n-manifold. In view of a myriad of (semi-)classical results, it seems a virtual certainty that the Lie algebra associated to the ordinary configuration spaces of M is (quasi-isomorphic to) the tensor product of the compactly supported Sullivan cochains of M with a free twisted Lie algebra on one generator in degree $n-1$ and weight 1 [**13, 16, 23, 32**]. It is likely possible to deduce this claim easily from either of the last two references with a little care taken in comparing models.

(3) *Orbit primitives.* The results of [**3**] suggest that the same description is valid verbatim for orbit configuration spaces, save that the phrase "free twisted Lie algebra" should be interpreted with respect to the category of equivariant bijections among free G-sets.

In more exotic settings, it is less clear what one should expect.

(4) *Generalized primitives.* What is the Lie algebra associated to generalized configuration spaces? Petersen's results in [**32**] suggest an expression in terms of the homology of the order complex associated to a collision structure. One imagines that the Chevalley–Eilenberg complex for this Lie algebra comprises the complexes described by Petersen, in the same way that the various Totaro spectral sequences [**38**] assemble into the Chevalley–Eilenberg complex of the Lie algebra described above in the case of ordinary configuration spaces [**22**].

One imagines that Theorem 1.3 is merely the beginning of a robust connection between stability phenomena and the twisted Chevalley–Eilenberg complex.

(5) *$\mathcal{FI}$-homology.* We show that sufficient connectivity of $L(X)$ implies representation stability. Is there a direct connection between this Lie algebra and $\mathcal{FI}$-(hyper)homology [**5**]?

(6) *Higher stability.* Is there an analogue of Theorem 1.3 characterizing higher order representation stability in terms of $L(X)$ [**30**]?

(7) *Beyond representation stability.* What finite generation properties does $L(X)$ dictate for projection spaces in more exotic contexts? What homological asymptotics do they imply? Pursuit of this direction almost certainly entails grappling with thorny Noetherianity problems for $\mathcal{C}$-twisted commutative algebras, already notoriously difficult (and mostly unsolved) in the classical case $\mathcal{C} = \mathcal{FB}$ [**31**].

It is likely that our framework and examples can be expanded considerably.

(8) *Local coefficients.* One imagines a theory in the vein of [**35**] treating the homology of projection spaces with compatibly twisted coefficients. Stability phenomena in the twisted (co)homology of configuration spaces is of considerable interest—see [**11**], for example.

(9) *Linear collision structures.* It is likely that a linear version of the theory developed in Section 3 would permit the expression of the homology of subspace arrangements as twisted Lie algebra homology.

(10) *Rational homotopy theory.* In the language of [**1**], the Day convolution and pointwise tensor products endow the category of presheaves of chain complexes with the structure of a 2-*monoidal category*, and the singular chains of a projection space carry the structure of a *double coalgebra* in this category, with the second comultiplication induced by the objectwise diagonal. Does this structure provide a faithful algebraic model for the homotopy theory of rationalized projection spaces [**33**]?

1.4. Conventions. Chain complexes are bounded below and homologically graded. The r-fold homological suspension is denoted $[r]$. When working with monoidal categories, we suppress the associator whenever possible. We use the symbols $\otimes$ and $\mathbb{1}$ for the tensor product and unit of a generic monoidal category, employing subscripts (rarely) to disambiguate as necessary (e.g., $\otimes_{\mathcal{C}}$). An additive tensor category is an additive category equipped with a symmetric monoidal structure whose tensor product distributes over finite coproducts. We write $\Gamma^k(V) = (V^{\otimes k})^{\Sigma_k}$, $\mathrm{Sym}^k(V) = V^{\otimes k}_{\Sigma_k}$, $\Gamma(V) = \bigoplus_{k\geq 0}\Gamma^k(V)$, and $\mathrm{Sym}(V) = \bigoplus_{k\geq 0}\mathrm{Sym}^k(V)$.

2. Projection categories

In this section, we introduce the projection category $\mathrm{Pr}(\mathcal{C})$ associated to a monoidal category $\mathcal{C}$. We then explore a selection of simple examples.

2.1. Definitions. Roughly speaking, we wish for a presheaf on the projection category $\mathrm{Pr}(\mathcal{C})$ to carry a structure map for each projection of an object onto a tensor factor. The following definition makes this idea precise.

DEFINITION 2.1. Let C_1 and C_2 be objects in the monoidal category $\mathcal{C}$.

(1) A *complementary morphism* from C_1 to C_2 is the data of an object D of $\mathcal{C}$ together with a morphism $f : C_1 \otimes D \to C_2$.
(2) The *composite* of the complementary morphisms $f_1 : C_1 \otimes D_1 \to C_2$ and $f_2 : C_2 \otimes D_2 \to C_3$ is the complementary morphism $f_2 \circ (f_1 \otimes D_2) : C_1 \otimes D_1 \otimes D_2 \to C_2$.
(3) An *elementary equivalence* from the complementary morphism $f : C_1 \otimes D \to C_2$ to the complementary morphism $f' : C_1 \otimes D' \to C_2$ is a map $g : D \to D'$ fitting into the commutative diagram

$$\begin{array}{ccc} C_1 \otimes D & \xrightarrow{C_1 \otimes g} & C_1 \otimes D' \\ & {\scriptstyle f}\searrow \quad \swarrow{\scriptstyle f'} & \\ & C_2 & \end{array}$$

(4) We say that two complementary morphisms are *equivalent* if they differ by a finite sequence of elementary equivalences.

We emphasize that the arrow g in the definition of an elementary equivalence is *not* required to be an isomorphism—see Remark 2.6 below for more on this point.

REMARK 2.2. The definition of a complementary morphism is biased by mapping order, i.e., whether to consider $C_1 \otimes D \to C_2$ or $C_1 \to C_2 \otimes D$. Up to opposites, the two choices produce essentially the same result. The definition is also biaed by tensor order, i.e., whether to consider $C_1 \otimes D$ or $D \otimes C_1$. Following this section, we will assume that $\mathcal{C}$ is symmetric monoidal, so this bias will play no role.

LEMMA 2.3. *Composition is associative and unital up to equivalence and well-defined on equivalence classes.*

PROOF. We give only an outline, leaving the (easy) details to the reader. For associativity, one uses the associator of $\mathcal{C}$ to produce an elementary equivalence between the two composites, and the compositional unit is the unitor $C \cong C \otimes \mathbb{1}$.

Well-definition in one composition factor follows from the commuting diagram

$$\begin{array}{ccccc} C_1 \otimes D_1 \otimes D_2' & \longrightarrow & C_2 \otimes D_2 & \longrightarrow & C_3 \\ \Big\downarrow{\scriptstyle C_1\otimes D_1\otimes g} & & \Big\downarrow{\scriptstyle C_2\otimes g} & & \Big\| \\ C_1 \otimes D_1 \otimes D_2 & \longrightarrow & C_2 \otimes D_2' & \longrightarrow & C_3, \end{array}$$

and a similar diagram establishes well-definition in the second factor. □

The lemma allows us our main definition.

DEFINITION 2.4. Let $\mathcal{C}$ be a monoidal category. The *projection category* of $\mathcal{C}$ is the category $\Pr(\mathcal{C})$ with the same objects as $\mathcal{C}$ and arrows the equivalence classes of complementary morphisms under composition.

The projection category is functorial for strong monoidal functors. Briefly, given such a functor $F : \mathcal{C} \to \mathcal{D}$, we define $\Pr(F) = F$ on objects, and we declare that $\Pr(F)$ send the equivalence class of the complementary morphism $C_1 \otimes D \to C_2$ to the equivalence class of the composite $F(C_1) \otimes F(D) \cong F(C_1 \otimes D) \to F(C_2)$. It is a simple matter to check that $\Pr(F)$ is a well-defined functor.

REMARK 2.5. The functoriality described above is a shadow of a larger structure. Specifically, the projection category $\Pr(\mathcal{C})$ is the truncation of an obvious bicategory, and an elaboration of the considerations of Section 4.1 shows that a lax monoidal functor between monoidal categories induces a pseudofunctor at the level of bicategories. We make no use of this extended functoriality.

REMARK 2.6. The projection category is closely related to a construction due to Quillen [**18**] (see also [**35**, p. 11]). More precisely, the projection category is obtained from the bicategory described in Remark 2.5 by replacing the morphism categories by their connected components, while Quillen's construction (in the case of a self-action) is obtained by replacing them with the connected components of their maximal subgroupoids—see Example 2.7 for a simple example illustrating the difference. In particular, one should expect the two to coincide only when $\mathcal{C}$ is itself a groupoid. The use of the full morphism category is essential to the universal property of Theorem 4.2—see Remark 4.6 and Example 4.7.

EXAMPLE 2.7. Let Δ^1 denote the walking arrow, i.e., the category with objects 0 and 1 and a unique non-identity morphism $e : 0 \to 1$. The usual rules of integer multiplication extend to a unique symmetric monoidal structure on Δ^1. With this monoidal structure, one finds that there are the exactly five complementary morphisms. Since Δ^1 has no non-identity isomorphisms, these five morphisms are distinct in Quillen's category. One checks that the resulting category is the walking retract, i.e., the category generated by the arrows $i : 0 \to 1$ and $r : 1 \to 0$ subject only to the relation $r \circ i = \mathrm{id}$, where $i = [0 \otimes 0 \xrightarrow{e} 1]$ and $r = [1 \otimes 0 \xrightarrow{\mathrm{id}} 0]$. In contrast, the commutative diagram

$$\begin{array}{ccc} 1 \otimes 0 \otimes 0 & \xrightarrow{\quad 1\otimes e \quad} & 1 \otimes 1 \\ & {\scriptstyle e}\searrow \quad \swarrow{\scriptstyle \mathrm{id}} & \\ & 1 & \end{array}$$

is a simple equivalence demonstrating the relation $i \circ r = \mathrm{id}$ in $\Pr(\Delta^1)$. Thus, the projection category is instead the walking isomorphism.

2.2. Groupoid examples. We begin with the primordial motivating example. Although this result is subsumed in Proposition 2.9, we include an independent proof, both for tactility and for later use.

PROPOSITION 2.8. *There is a canonical isomorphism of categories*

$$\mathrm{Pr}(\mathcal{FB}) \cong \mathcal{FI}.$$

PROOF. The two categories have the same objects. We define a functor from left to right on arrows by sending the equivalence class of the complementary morphism $f : I_1 \sqcup J \cong I_2$ to the injection $f|_{I_1}$. One checks immediately that this prescription is invariant under equivalence and respects identities. As for composition, the complementary morphisms $f : I_1 \sqcup J_1 \cong I_2$ and $g : I_2 \sqcup J_2 \cong I_3$ compose to give $g \circ (f \sqcup J_2) : I_1 \sqcup J_1 \sqcup J_2 \to I_3$, and

$$(g \circ (f \sqcup J_2)) \,|_{I_1} = g \circ (f \sqcup J_2) \,|_{I_1} = g|_{I_2} \circ f|_{I_1},$$

as required. To define the functor from right to left, we send the injection $i : I_1 \to I_2$ to the equivalence class of the complementary morphism $I_1 \sqcup (I_2 \setminus \mathrm{im}(i)) \cong I_2$. The injection $I = I$ is sent to the equivalence class of $I \sqcup \varnothing \cong I$, which is the identity of I in $\mathrm{Pr}(\mathcal{FB})$, and the composite of $i : I_1 \to I_2$ and $j : I_2 \to I_3$ is sent to the equivalence class of $I_1 \sqcup (I_3 \setminus \mathrm{im}(j \circ i)) \cong I_3$, which is equivalent to the appropriate composite via the bijection

$$I_2 \setminus \mathrm{im}(i) \sqcup I_3 \setminus \mathrm{im}(j) \cong I_3 \setminus \mathrm{im}(j \circ i)$$

induced by j.

We have shown that both assignments are functors, and the composite functors are the respective identities on objects by construction. On arrows, given $f : I_1 \sqcup J \cong I_2$, we have

$$I_2 \setminus \mathrm{im}(f|_{I_1}) = f(J),$$

and $f^{-1}|_{f(J)}$ provides an equivalence between our original complementary morphism and the complementary morphism $I_1 \sqcup f(J) \cong I_2$, which represents its image under the composite functor. We leave it to the reader to check that the other composite is also the identity on arrows. □

We now consider a class of example recurrent throughout the study of stability phenomena. In the next result, the assumption that $\mathcal{C}$ be skeletal is made solely for simplicity of exposition.

PROPOSITION 2.9. *Let $\mathcal{C}$ be a skeletal symmetric monoidal groupoid. There is a canonical bijection*

$$\mathrm{Hom}_{\mathrm{Pr}(\mathcal{C})}(C_1, C_2) \cong \coprod_{C_2 = C_1 \otimes D} \mathrm{Aut}(C_2)/\mathrm{Aut}(D)$$

under which composition is given by the dashed filler in the following commutative diagram:

$$\begin{array}{ccccc} \mathrm{Aut}(C_2) \times \mathrm{Aut}(C_3) & \xrightarrow{(-\otimes D_2)\times \mathrm{id}} & \mathrm{Aut}(C_3) \times \mathrm{Aut}(C_3) & \xrightarrow{\circ} & \mathrm{Aut}(C_3) \\ \downarrow & & & & \downarrow \\ \mathrm{Aut}(C_2)/\mathrm{Aut}(D_1) \times \mathrm{Aut}(C_3)/\mathrm{Aut}(D_2) & & \dashrightarrow & & \mathrm{Aut}(C_3)/\mathrm{Aut}(D_1 \otimes D_2). \end{array}$$

Proof. For each object D such that $C_2 = C_1 \otimes D$, an automorphism of C_2 determines a complementary morphism $C_1 \otimes D \cong C_2$, and every complementary morphism is determined in this way. Thus, we have a canonical surjection

$$\coprod_{C_2 = C_1 \otimes D} \operatorname{Aut}(C_2) \to \operatorname{Hom}_{\operatorname{Pr}(\mathcal{C})}(C_1, C_2).$$

By definition, two automorphisms of C_2 differ by an automorphism of the form $C_1 \otimes g$ for $g \in \operatorname{Aut}(D)$ if and only if the corresponding complementary morphisms are equivalent. It follows that the above surjection descends to the indicated disjoint union of orbit sets, and that each of the resulting functions $\operatorname{Aut}(C_2)/\operatorname{Aut}(D) \to \operatorname{Hom}_{\operatorname{Pr}(\mathcal{C})}(C_1, C_2)$ is injective. Since $\mathcal{C}$ is skeletal, complementary morphisms indexed by different choices of D are never equivalent, so the function as a whole is injective, implying the first claim. The (essentially immediate) verification of the second claim is left to the reader. $\square$

In the following result, $\mathcal{FI}_G$ denotes the category of equivariant injections among free G-sets with finitely many orbits, where G is a fixed group, and $\mathcal{VI}_{\mathbb{F}}$ denotes the category of linear injections among finite dimensional vector spaces over a fixed vector space $\mathbb{F}$ (resp. $\mathcal{FB}_G$, $\mathcal{VB}_{\mathbb{F}}$, bijections).

Corollary 2.10. *Let G be a group and $\mathbb{F}$ a field. There are the following canonical isomorphisms of categories:*

$$\operatorname{Pr}(\mathcal{FB}_G) \cong \mathcal{FI}_G$$
$$\operatorname{Pr}(\mathcal{VB}_{\mathbb{F}}) \cong \mathcal{VI}_{\mathbb{F}}.$$

2.3. The projection category of a deformation. We close this section with an observation on the relationship between the formation of projection categories and subcategories of a certain type.

Definition 2.11. A subcategory $\iota : \mathcal{C}_0 \subseteq \mathcal{C}$ is a *deformation* of $\mathcal{C}$ if there is a functor $R : \mathcal{C} \to \mathcal{C}_0$ and a natural transformation $\tau : \iota \circ R \to \operatorname{id}$.

Note that any subcategory containing a deformation is itself a deformation.

Proposition 2.12. *Let $\mathcal{C}$ be a monoidal category and $\mathcal{C}_0 \subseteq \mathcal{C}$ a full subcategory containing $\mathbb{1}$ and closed under tensor products. If $\mathcal{C}_0$ is a deformation of $\mathcal{C}$, then the induced functor $\operatorname{Pr}(\mathcal{C}_0) \to \operatorname{Pr}(\mathcal{C})$ is fully faithful with image the subcategory $\operatorname{Pr}(\mathcal{C})_0 \subseteq \operatorname{Pr}(\mathcal{C})$ with objects the objects of $\mathcal{C}_0$ and morphisms the equivalence classes of complementary morphisms $f : C_1 \otimes D \to C_2$ such that D lies in $\mathcal{C}_0$.*

Proof. Prerequisitely, we note that $\operatorname{Pr}(\mathcal{C})_0$ is, in fact, a subcategory by closure under tensor products. This same assumption, together with fullness and the assumption that $\mathbb{1} \in \mathcal{C}_0$, implies that the tensor product of $\mathcal{C}$ restricts to a monoidal structure on $\mathcal{C}_0$ with the same unit and coherence isomorphisms, so $\operatorname{Pr}(\mathcal{C}_0)$ is defined, and the inclusion $\mathcal{C}_0 \subseteq \mathcal{C}$ is strong monoidal, so $\operatorname{Pr}(\mathcal{C}_0) \to \operatorname{Pr}(\mathcal{C})$ is defined. We claim that this functor factors through the inclusion of $\operatorname{Pr}(\mathcal{C})_0$ as an isomorphism. The existence, surjectivity on objects, and fullness of the factorization being essentially immediate, the main point is to verify faithfulness. Consider the zig-zag of simple equivalences of the form

$$\begin{array}{ccccc} C_1 \otimes D & \xrightarrow{C_1 \otimes g} & C_1 \otimes D' & \xleftarrow{C_1 \otimes g'} & C_1 \otimes D'' \\ & {\scriptstyle f} \searrow & \downarrow {\scriptstyle f'} & \swarrow {\scriptstyle f''} & \\ & & C_2, & & \end{array}$$

with all objects but D' lying in $\mathcal{C}_0$. By assumption, there is a functor $R : \mathcal{C} \to \mathcal{C}_0$ and a natural transformation $\tau : \iota \circ R \to \mathrm{id}$, from which we derive the enlarged diagram

$$\begin{array}{ccccc} C_1 \otimes R(D) & \xrightarrow{C_1 \otimes R(g)} & C_1 \otimes R(D') & \xleftarrow{C_1 \otimes R(g')} & C_1 \otimes R(D'') \\ {\scriptstyle C_1 \otimes \tau} \downarrow & & \downarrow {\scriptstyle C_1 \otimes \tau} & & \downarrow {\scriptstyle C_1 \otimes \tau} \\ C_1 \otimes D & \xrightarrow{C_1 \otimes g} & C_1 \otimes D' & \xleftarrow{C_1 \otimes g'} & C_1 \otimes D'' \\ & {\scriptstyle f} \searrow & \downarrow {\scriptstyle f'} & \swarrow {\scriptstyle f''} & \\ & & C_2. & & \end{array}$$

This diagram supplies the following chain of simple equivalences:

$$f \sim f \circ C_1 \otimes \tau \sim f' \circ C_1 \otimes \tau \sim f'' \circ C_1 \otimes \tau \sim f'',$$

all of which lie in $\mathcal{C}_0$. In the same way, an arbitrary zig-zag of simple equivalences expressing the equivalence in $\mathcal{C}$ of two complementary morphisms with source and target in $\mathcal{C}_0$ may be replaced by a (perhaps longer) zig-zag of simple equivalences in $\mathcal{C}_0$. In other words, complementary morphisms in $\mathcal{C}_0$ are equivalent if and only if they are equivalent as complementary morphisms in $\mathcal{C}$, which is faithfulness. □

3. Collision structures

In this section, we develop a general combinatorial framework encompassing all of our examples. Inspired by [**32**], we define a *collision structure* to be a set of partitions of a finite set closed under merging blocks. We define a notion of morphism between collision structures and calculate the projection category associated to the category of collision structures and bijections (Theorem 3.11).

3.1. Definitions. We begin by recalling a few standard ideas regarding partitions.

DEFINITION 3.1. Let I be a set. A *partition* of I is a set P of nonempty subsets of I, called *blocks*, such that every element of I is contained in exactly one block of P. We say that P is a *refinement* of P', written $P \leq P'$, if every block of P' is a union of blocks of P.

The data of a partition P of I is equivalent to the data of the equivalence relation $\sim_P$ on I given by declaring that two elements are equivalent provided they lie in the same block of P. Note that the empty set admits a unique partition, which is itself empty.

Partitions of I form a poset Π_I under refinement, which has the unique minimum $\{\{i\}\}_{i \in I}$ and, if I is nonempty, the unique maximum $\{I\}$.

DEFINITION 3.2. A *collision structure* on the set I is an upward closed subset $S \subseteq \Pi_I$, i.e., such that $P' \in S$ whenever $P \in S$ and $P \leq P'$.

We think of a collision structure as prescribing which collisions among elements of the underlying set are forbidden. From this point of view, the requirement of upward closure is obvious.

The intersection or union of collision structures on a fixed set is again a collision structure; therefore, it is sensible to speak of the largest and smallest collision structure with a given property, and of the collision structure generated by a set of partitions.

EXAMPLE 3.3. The *trivial* collision structure $\varnothing \subseteq \Pi_I$ is initial among collision structures on I.

Given a partition P of I and an injection $f : I \to J$, we write f_*P for the partition of J given by the images under f of the blocks of P, together with the singletons in the complement of the image of f. Given a subset $S \subseteq \Pi_I$, we write f_*S for the collision structure generated by the set $\{f_*P : P \in S\}$.

DEFINITION 3.4. Let S and T be collision structures on I and J. We say that an injection $f : I \to J$ is a *map of collision structures* if $f_*S \subseteq T$.

Since we clearly have $g_*f_*S = (g \circ f)_*S$, maps of collision structures on finite sets form a category $\mathcal{CSI}$. We will also be interested in the wide subcategory $\mathcal{CSB}$ of bijective maps of collision structures.

As we will see below in Section 6.1, a collision structure S has an associated configuration space, in which collisions among particles are forbidden if the resulting partition lies in S. These spaces are precisely the generalized configuration spaces of [**32**]. We close this section by observing that the category of collision structures subsumes the combinatorics underlying the graphical configuration spaces of [**10**] and the simplicial configuration spaces of [**7**]

EXAMPLE 3.5. A graph Γ determines a collision structure S_Γ on its set V of vertices by declaring that $P \in \Pi_V$ lies in S_Γ if and only if some block of P contains an edge. An injection between vertex sets is a map of collision structures if and only if it is a graph homomorphism.

EXAMPLE 3.6. A simplicial complex K determines a collision structure S_K on its set V of vertices by declaring that $P \in \Pi_V$ lies in S_K if and only if some block of P is not a simplex of K (we use that the set of simplices is closed under the formation of subsets). An injection between vertex sets is a map of collision structures if and only if it preserves non-simplices.[1]

Note that the collision structure associated to a graph according to Example 3.5 coincides with the collision structure associated to its independence complex according to Example 3.6.

From these examples, we obtain the categories $\mathcal{GI}$ and $\mathcal{SCI}$ of injections among graphs and simplicial complexes, respectively, both full subcategories of $\mathcal{CSI}$, and similarly for bijections.

[1]Maps of this kind are called "cosimplicial" in [**7**]. We avoid this terminology for reasons that are likely obvious.

3.2. Monoidal structure and projection category. The main result of this section identifies the projection category of the category $\mathcal{CSB}$ of collision structures and bijections. In order to formulate such a result, we first require a monoidal structure.

DEFINITION 3.7. Let S and T be collision structures on I and J, respectively. The *disjoint union* of S and T is the collision structure $S \sqcup T \subseteq \Pi_{I \sqcup J}$ generated by $(\iota_I)_*S \cup (\iota_J)_*T$, where $\iota_I : I \to I \sqcup J$ denotes the inclusion (resp. J).

This definition is arranged so that ι_I and ι_J are maps of collision structures.

LEMMA 3.8. *Disjoint union of collision structures extends to a unique symmetric monoidal structure such that the forgetful functor* $\mathcal{CSB} \to \mathcal{FB}$ *is strong monoidal.*

PROOF. The main point is to verify that the (suppressed) associator of finite sets is an isomorphism of collision structures, but it is not difficult to check, given collision structures S_i on I_i for $i \in \{1,2,3\}$, that $(S_1 \sqcup S_2) \sqcup S_3$ and $S_1 \sqcup (S_2 \sqcup S_3)$ are both generated by the set $\bigcup_{i=1}^{3} (\iota_{I_i})_* S_i$. □

By functoriality for strong monoidal functors, we obtain an induced functor at the level of projection categories, concerning which we have the following result.

LEMMA 3.9. *The induced functor* $\mathrm{Pr}(\mathcal{CSB}) \to \mathrm{Pr}(\mathcal{FB})$ *is faithful.*

PROOF. For $i \in \{1,2\}$, let S_i and T_i be collision structures on I_i and J_i, respectively. Consider the following diagram of bijections:

$$\begin{array}{ccc} I_1 \sqcup J_1 & \xrightarrow{I_1 \sqcup g} & I_1 \sqcup J_2 \\ & {\scriptstyle f_1} \searrow \quad \swarrow {\scriptstyle f_2} & \\ & I_2. & \end{array}$$

It suffices to show, assuming that f_1 and f_2 are maps of collision structures (and g not necessarily so), that f_1 and f_2 are equivalent as complementary morphisms in $\mathcal{CSB}$. Let T_3 denote the collision structure on J_2 generated by g_*T_1 and T_2. Then g is a map of collision structures from T_1 to T_3, and the identity is a map of collision structures from T_2 to T_3. We claim that f_2 is a map of collision structures from $S_1 \sqcup T_3$ to S_2, for which it suffices to verify the following three containments:

$$\begin{aligned} (f_2)_*(\iota_{I_1})_* S_1 &\subseteq S_2 \\ (f_2)_*(\iota_{J_2})_* g_* T_1 &\subseteq S_2 \\ (f_2)_*(\iota_{J_2})_* T_2 &\subseteq S_2. \end{aligned}$$

The first and third containment are the assumption that f_2 is a map of collision structures, and the second follows from the assumption that f_1 is so, since $f_2 \circ \iota_{J_2} \circ g = f_1 \circ \iota_{J_1}$ by commutativity. We have established the existence of the commutative diagram

$$\begin{array}{ccccc} S_1 \sqcup T_1 & \xrightarrow{I_1 \sqcup g} & S_1 \sqcup T_3 & \xleftarrow{\mathrm{id}} & S_1 \sqcup T_2 \\ & {\scriptstyle f_1} \searrow & \downarrow {\scriptstyle f_2} & \swarrow {\scriptstyle f_2} & \\ & & I_2 & & \end{array}$$

of collision structures, which represents a pair of elementary equivalences connecting f_1 and f_2. $\square$

EXAMPLE 3.10. Via Examples 3.5 and 3.6, the disjoint union of graphs and of simplicial complexes corresponds to the disjoint union of collision structures. In other words, the inclusions $\mathcal{GB} \subseteq \mathcal{CSB}$ and $\mathcal{SCB} \subseteq \mathcal{CSB}$ are strong monoidal.

We come now to the main result of this section.

THEOREM 3.11. *There is a canonical isomorphism of categories*

$$\mathrm{Pr}(\mathcal{CSB}) \cong \mathcal{CSI}.$$

PROOF. The two categories have the same objects, and we wish to extend this equality to an isomorphism of categories. Forgetting collision structures determines the vertical functors in the diagram

$$\begin{array}{ccc} \mathrm{Pr}(\mathcal{CSB}) & \dashrightarrow & \mathcal{CSI} \\ \downarrow & & \downarrow \\ \mathrm{Pr}(\mathcal{FB}) & \xrightarrow{\cong} & \mathcal{FI}, \end{array}$$

where the bottom functor is the isomorphism of Proposition 2.8. Since the vertical functors are faithful by Lemma 3.9 and by definition, respectively, the dashed functor (extending the identity on objects) is unique if it exists.

By definition, an arrow from S_1 to S_2 in the source of the putative functor is represented by a bijection $f : I_1 \sqcup J \cong I_2$ such that $f_*(S_1 \sqcup T) \subseteq S_2$. We have

$$(f|_{I_1})_* S_1 = (f \circ \iota_{I_1})_* S_1 = f_*(\iota_{I_1})_* S_1 \subseteq f_*(S_1 \sqcup T) \subseteq S_2,$$

so the injection $f|_{I_1}$ defines a morphism from S_1 to S_2 in $\mathcal{CSI}$. Thus, the dashed functor exists. To obtain its inverse, it suffices for the same reasons to note, given an injection $i : I_1 \to I_2$ with $i_* S_1 \subseteq S_2$, that the bijection $I_1 \sqcup (I_2 \setminus \mathrm{im}(i)) \cong I_2$ lifts uniquely to a complementary morphism in $\mathcal{CSB}$. For existence, we note that the bijection in question is a map of collision structures when $I_2 \setminus \mathrm{im}(i)$ carries the empty collision structure. For uniqueness, we note that, for any collision structure T on $I_2 \setminus \mathrm{im}(i)$, any permutation of this set defines a morphism of collision structures $\varnothing \to T$; therefore, any lift of $I_1 \sqcup (I_2 \setminus \mathrm{im}(i)) \cong I_2$ to a complementary morphism is equivalent to the lift constructed above. $\square$

COROLLARY 3.12. *There are the following canonical isomorphisms of categories:*

$$\mathrm{Pr}(\mathcal{GB}) \cong \mathcal{GI}$$
$$\mathrm{Pr}(\mathcal{SCB}) \cong \mathcal{SCI}.$$

PROOF. We aim to use Proposition 2.12, the main point being to verify that $\mathcal{GB}$ and $\mathcal{SCB}$ are deformations of $\mathcal{CSB}$. We begin by observing that the assignment of a finite set to its trivial collision structure defines a fully faithful functor $\mathcal{FB} \to \mathcal{CSB}$, which is a section of the forgetful functor. Identifying $\mathcal{FB}$ with its essential image under this functor, we observe that $\mathcal{FB}$ is a deformation of $\mathcal{CSB}$, since the trivial collision structure is initial. Moreover, we have the containment $\mathcal{FB} \subseteq \mathcal{GB} \cap \mathcal{SCB}$; indeed, the trivial collision structure on I corresponds to the discrete graph on I and to the simplex spanned by I. Since a subcategory containing a deformation is itself a deformation, the proof is complete. $\square$

4. Monoidal matters

In this section, we consider interactions among various tensor products. First, we prove the structural result that the opposite of the projection category represents oplax symmetric monoidal functors with Cartesian target (Corollary 4.3). Second, we show that, under restrictive but applicable hypotheses, a variant of Day convolution identifies such oplax functors with certain coalgebras (Corollary 4.12). These two results form the bridge from combinatorial structure to algebraic structure underpinning our proof of Theorem 1.2.

4.1. Projection categories and lax structures. The goal of this section is to establish a universal property of the projection category in the symmetric setting. Before articulating this universal property in Theorem 4.2 below, we pause to establish notation regarding a few standard concepts.

DEFINITION 4.1. Let $\mathcal{C}$ and $\mathcal{D}$ be monoidal categories.

(1) A *lax (monoidal)* structure on a functor $F : \mathcal{C} \to \mathcal{D}$ is a natural transformation $\mu : F \otimes F \to F \circ \otimes$ and a morphism $\eta : \mathbb{1} \to F(\mathbb{1})$ such that the following diagrams commute for all $C_1, C_2, C_3 \in \mathcal{C}$.

$$\begin{array}{ccc}
F(C_1 \otimes (C_2 \otimes C_3)) & \overset{\sim}{=\!=\!=} & F((C_1 \otimes C_2) \otimes C_3) \\
{\scriptstyle \mu}\big\uparrow & & \big\uparrow{\scriptstyle \mu} \\
F(C_1) \otimes F(C_2 \otimes C_3) & & F(C_1 \otimes C_2) \otimes F(C_3) \\
{\scriptstyle F(C_1)\otimes\mu}\big\uparrow & & \big\uparrow{\scriptstyle \mu\otimes F(C_3)} \\
F(C_1) \otimes (F(C_2) \otimes F(C_3)) & \overset{\sim}{=\!=\!=} & (F(C_1) \otimes F(C_2)) \otimes F(C_3)
\end{array}$$

$$\begin{array}{ccc}
F(C_1 \otimes \mathbb{1}) & \xleftarrow{\ \mu\ } & F(C_1) \otimes F(\mathbb{1}) \\
{\scriptstyle \wr}\big\| & & \big\uparrow{\scriptstyle F(C_1)\otimes\eta} \\
F(C_1) & \overset{\sim}{=\!=\!=} & F(C_1) \otimes \mathbb{1}.
\end{array}$$

A *lax (monoidal) functor* is a functor equipped with a (typically suppressed) lax structure.

(2) Suppose that $\mathcal{C}$ and $\mathcal{D}$ are symmetric monoidal. We say that the lax functor $F : \mathcal{C} \to \mathcal{D}$ is *symmetric* if the following diagram commutes:

$$\begin{array}{ccc}
F(C_1 \otimes C_2) & \overset{\sim}{=\!=\!=} & F(C_2 \otimes C_1) \\
{\scriptstyle \mu}\big\uparrow & & \big\uparrow{\scriptstyle \mu} \\
F(C_1) \otimes F(C_2) & \overset{\sim}{=\!=\!=} & F(C_2) \otimes F(C_1).
\end{array}$$

(3) Let F_1 and F_2 be lax functors. A natural transformation $\tau : F_1 \to F_2$ is *monoidal* if the following diagrams commute:

$$\begin{array}{ccc}
F_1(C_1 \otimes C_2) & \xrightarrow{\ \tau\ } & F_2(C_1 \otimes C_2) \\
{\scriptstyle \mu}\big\uparrow & & \big\uparrow{\scriptstyle \mu} \\
F_1(C_1) \otimes F_1(C_2) & \xrightarrow{\ \tau\otimes\tau\ } & F_2(C_1) \otimes F_2(C_2)
\end{array}
\qquad
\begin{array}{ccc}
F_1(\mathbb{1}) & \xrightarrow{\ \tau\ } & F_2(\mathbb{1}) \\
{\scriptstyle \eta}\nwarrow & & \nearrow{\scriptstyle \eta} \\
& \mathbb{1} &
\end{array}$$

(4) A (symmetric) *oplax* structure on $F : \mathcal{C} \to \mathcal{D}$ is a (symmetric) lax structure on F^{op}.

It is immediate from the definitions that monoidality and symmetry is closed under composition; therefore, lax symmetric monoidal functors and monoidal natural transformations form a category $\mathrm{Fun}^{\mathrm{lax}}(\mathcal{C}, \mathcal{D})$. Likewise, we have the category $\mathrm{Fun}^{\mathrm{oplax}}(\mathcal{C}, \mathcal{D}) = \mathrm{Fun}^{\mathrm{lax}}(\mathcal{C}^{\mathrm{op}}, \mathcal{D}^{\mathrm{op}})^{\mathrm{op}}$ of oplax symmetric monoidal functors. The reader should note that we do not reflect the condition of symmetry in the notation, since we will have no cause to consider nonsymmetric (op)lax functors.

THEOREM 4.2. *Let $\mathcal{C}$ be a symmetric monoidal category and $\mathcal{D}$ a category with finite coproducts. There is a canonical isomorphism of categories*

$$\mathrm{Fun}(\mathrm{Pr}(\mathcal{C}), \mathcal{D}) \cong \mathrm{Fun}^{\mathrm{lax}}(\mathcal{C}, \mathcal{D}).$$

In our application, it will be the dual version of this result that will be of most interest.

COROLLARY 4.3. *Let $\mathcal{C}$ be a symmetric monoidal category and $\mathcal{D}$ a category with finite products. There is a canonical isomorphism of categories*

$$\mathrm{Fun}(\mathrm{Pr}(\mathcal{C})^{\mathrm{op}}, \mathcal{D}) \cong \mathrm{Fun}^{\mathrm{oplax}}(\mathcal{C}^{\mathrm{op}}, \mathcal{D}).$$

PROOF. By our assumption on $\mathcal{D}$, the opposite category $\mathcal{D}^{\mathrm{op}}$ has finite coproducts, so Theorem 4.2 supplies the middle of the three isomorphisms of categories:

$$\mathrm{Fun}(\mathrm{Pr}(\mathcal{C})^{\mathrm{op}}, \mathcal{D}) \cong \mathrm{Fun}(\mathrm{Pr}(\mathcal{C}), \mathcal{D}^{\mathrm{op}})^{\mathrm{op}} \cong \mathrm{Fun}^{\mathrm{lax}}(\mathcal{C}, \mathcal{D}^{\mathrm{op}})^{\mathrm{op}} \cong \mathrm{Fun}^{\mathrm{oplax}}(\mathcal{C}^{\mathrm{op}}, \mathcal{D}).$$

□

We begin the proof of the theorem by observing that $\mathcal{C}$ is the source of a natural functor to its own projection category.

LEMMA 4.4. *The assignments $\iota(C) = C$ and $\iota(C_1 \to C_2) = [C_1 \otimes \mathbb{1} \cong C_1 \to C_2]$ determine a functor $\iota : \mathcal{C} \to \mathrm{Pr}(\mathcal{C})$.*

PROOF. It is immediate that ι preserves identities. For composition, we appeal to the commutative diagram

$$\begin{array}{ccccccc}
C_1 \otimes \mathbb{1} \otimes \mathbb{1} & \overset{\sim}{=\!=} & C_1 \otimes \mathbb{1} & \longrightarrow & C_2 \otimes \mathbb{1} & \overset{\sim}{=\!=} & C_2 \\
\wr \| & & \wr \| & & \wr \| & & \downarrow \\
C_1 \otimes \mathbb{1} & \overset{\sim}{=\!=} & C_1 & \longrightarrow & C_2 & \longrightarrow & C_3
\end{array}$$

The composite in the bottom row represents $\iota(C_1 \to C_2 \to C_3)$, while the full clockwise composite represents $\iota(C_2 \to C_3) \circ \iota(C_1 \to C_2)$. Since the leftmost vertical arrow is the identity of C_1 tensored with the unitor, commutativity implies that the two complementary morphisms are equivalent, as desired. □

The equivalence of Theorem 4.2 will be given by restriction along the functor ι of Lemma 4.4. The following result will allow us to make sense of this idea.

LEMMA 4.5. *Let $\mathcal{D}$ be category with finite coproducts.*

(1) *Given a functor $G : \mathrm{Pr}(\mathcal{C}) \to \mathcal{D}$, the restriction $\iota^* G$ carries a canonical lax structure, which is symmetric.*
(2) *Given a natural transformation $\tau : G_1 \to G_2$ in $\mathrm{Fun}(\mathrm{Pr}(\mathcal{C}), \mathcal{D})$, the restriction $\iota^* \tau$ is monoidal.*

PROOF. Since the unit in $\mathcal{D}$ is initial, we have the unique map $\mathbb{1} \to G(\mathbb{1})$. The identity and the symmetry of $\mathcal{C}$ determine complementary morphisms from C_1 and C_2 to $C_1 \otimes C_2$, respectively. Applying G and invoking the universal property of the coproduct in $\mathcal{D}$, we obtain a family of maps of the form $G(C_1) \sqcup G(C_2) \to G(C_1 \otimes C_2)$. The verification that these maps form a natural transformation, although straightforward, reveals a subtle asymmetry, which may at first be surprising; for this reason, we choose to include some details. The verification of the axioms of a symmetric lax structure and a monoidal natural transformation are left to the enthusiastic reader.

By the universal property of the coproduct, naturality in the first variable amounts to the commutativity of the diagrams obtained by applying G to the following two diagrams in $\mathrm{Pr}(\mathcal{C})$:

$$\begin{array}{ccc} C_1 & \xrightarrow{[C_1\otimes\mathbb{1}\cong C_1\xrightarrow{f}C_1']} & C_1' \\ {\scriptstyle [C_1\otimes C_2=C_1\otimes C_2]}\big\downarrow & & \big\downarrow{\scriptstyle [C_1'\otimes C_2=C_1'\otimes C_2]} \\ C_1\otimes C_2 & \xrightarrow{[C_1\otimes C_2\otimes\mathbb{1}\cong C_1\otimes C_2\xrightarrow{f\otimes C_2}C_1'\otimes C_2]} & C_1'\otimes C_2. \end{array}$$

$$\begin{array}{ccccc} & & C_2 & & \\ & \swarrow{\scriptstyle [C_2\otimes C_1\cong C_1\otimes C_2]} & & \searrow{\scriptstyle [C_2\otimes C_1'\cong C_1'\otimes C_2]} & \\ C_1\otimes C_2 & & \xrightarrow{[C_1\otimes C_2\otimes\mathbb{1}\cong C_1\otimes C_2\xrightarrow{f\otimes C_2}C_1'\otimes C_2]} & & C_1'\otimes C_2 \end{array}$$

The composites in the first diagram are represented by the respective rows of the following commutative diagram in $\mathcal{C}$:

$$\begin{array}{ccccc} C_1\otimes\mathbb{1}\otimes C_2 & \overset{\sim}{=\!=} & C_1\otimes C_2 & \xrightarrow{f\otimes C_2} & C_1'\otimes C_2 \\ \wr\Vert & & \Vert & & \Vert \\ C_1\otimes C_2\otimes\mathbb{1} & \overset{\sim}{=\!=} & C_1\otimes C_2 & \xrightarrow{f\otimes C_2} & C_1'\otimes C_2. \end{array}$$

As indicated, the symmetry of $\mathcal{C}$ supplies an elementary equivalence between the two, establishing the desired commutativity. On other hand, the counterclockwise composite and righthand diagonal arrow in the second diagram are represented by the clockwise composite and bottom arrow, respectively, of the following commutative diagram in $\mathcal{C}$:

$$\begin{array}{ccc} C_2\otimes C_1\otimes\mathbb{1} & \overset{\sim}{=\!=} & C_1\otimes C_2\otimes\mathbb{1} \\ \wr\Vert & & \Vert\wr \\ C_2\otimes C_1 & \overset{\sim}{=\!=} & C_1\otimes C_2 \\ {\scriptstyle C_2\otimes f}\big\downarrow & & \big\downarrow{\scriptstyle f\otimes C_2} \\ C_2\otimes C_1' & \overset{\sim}{=\!=} & C_1'\otimes C_2. \end{array}$$

The lefthand vertical composite gives a simple equivalence between the two, establishing the claim. Naturality in the second variable is similar. □

Note that the arrow providing the final simple equivlanece of this proof is in general *not an isomorphism*—see Remark 4.6 below.

Proof of Theorem 4.2. Invoking Lemma 4.5, we obtain the (abusively named) functor

$$\iota^* : \mathrm{Fun}(\mathrm{Pr}(\mathcal{C}), \mathcal{D}) \to \mathrm{Fun}^{\mathrm{lax}}(\mathcal{C}, \mathcal{D}),$$

which we claim to be an isomorphism. To prove this claim, we construct an inverse isomorphism explicitly. Given a symmetric lax functor $F : \mathcal{C} \to \mathcal{D}$, we define the value of $\overline{F} : \mathrm{Pr}(\mathcal{C}) \to \mathcal{D}$ on the equivalence class of $f : C_1 \otimes D \to C_2$ to be the composite

$$F(C_1) \to F(C_1) \sqcup F(D) \xrightarrow{\mu} F(C_1 \otimes D) \xrightarrow{F(f)} F(C_2).$$

One checks immediately that this definition is independent of the choice of representative and that $\overline{F}$, so defined, is a functor. To conclude, it suffices to verify that the objectwise identities $\overline{\iota^* G}(C) = G(C)$ and $\iota^*\overline{F}(C) = F(C)$ form a natural transformation and a monoidal natural transformation, respectively.

Unpacking the definitions, one finds that the value of $\overline{\iota^* G}$ on the equivalence class represented by $C_1 \otimes D \to C_2$ is the value of G on the equivalence class of the composite of $C_1 \otimes D \otimes \mathbb{1} \cong C_1 \otimes D \to C_2$, viewed as a complementary morphism with source $C_2 \otimes D$, with $C_1 \otimes D = C_1 \otimes D$, viewed as a complementary morphism with source C_1. This composite is simply $C_1 \otimes D \otimes \mathbb{1} \cong C_1 \otimes D \to C_2$, viewed as a complementary morphism with source C_1, which is equivalent to our original complementary morphism via the unitor $D \otimes \mathbb{1} \cong D$. It follows that $\overline{\iota^* G} = G$.

Similarly, one finds that the value of $\iota^*\overline{F}$ on a morphism $C_1 \to C_2$ is the composite in the top row of the diagram

$$\begin{array}{ccccccccc} F(C_1) & \longrightarrow & F(C_1)\sqcup F(\mathbb{1}) & \xrightarrow{\mu} & F(C_1\otimes\mathbb{1}) & \overset{\sim}{=\!=} & F(C_1) & \longrightarrow & F(C_2)\\ & \overset{\sim}{\diagdown\!\!\diagdown} & \uparrow{\scriptstyle F(C_1)\sqcup\eta} & & \overset{\sim}{\diagup\!\!\diagup} & & & & \\ & & F(C_1)\sqcup\mathbb{1} & & & & & & \end{array}$$

Since the diagram commutes by our assumption on F, this composite is simply the value of F on our original morphism. It follows that $\iota^*\overline{F} = F$ as bare functors, so it remains to verify that the two lax structures coincide. Since the unit in $\mathcal{D}$ is initial, the second diagram in the definition of a monoidal natural transformation commutes automatically. As for the first, it is an easy exercise to check that the lax structure morphism for $\iota^*\overline{F}$ produced by Lemma 4.5, as a morphism $F(C_1) \sqcup F(C_2) \to F(C_1 \otimes C_2)$ in $\mathcal{D}$, has the same components as μ; we remark only that the argument uses the compatibility of μ with the respective monoidal symmetries. The claim then follows from the universal property of the coproduct in $\mathcal{D}$. □

Remark 4.6. As our proof shows, the (rather curious) situation is that the universal property of Theorem 4.2 fails to hold for Quillen's category not because one cannot define the desired lax monoidal structure maps, but rather because they do not form a natural transformation unless $\mathcal{C}$ is a groupoid. The coarser equivalence relation of $\mathrm{Pr}(\mathcal{C})$ is precisely what is needed to repair this defect.

Example 4.7. We return to Example 2.7. By definition, the data of a functor from Δ^1 is that of an arrow $f : X \to Y$ in the target, and our previous analysis shows that the data of an extension of this functor to Quillen's construction is that of a retraction g of f. Assuming the target category to have finite coproducts, suppose that we are instead given the data of a lax monoidal structure on the functor in question. The composite arrow

$$g : Y \to X \sqcup Y \to X$$

is again a left inverse to f, as is readily seen from the commutative diagram

$$\begin{array}{ccccc} X \sqcup Y & \xrightarrow{f \sqcup Y} & Y \sqcup Y & \xleftarrow{! \sqcup Y} & \varnothing \sqcup Y \\ \downarrow & & \downarrow & \nearrow_{\cong} & \\ X & \xrightarrow{f} & Y. & & \end{array}$$

According to Example 2.7, the category $\mathrm{Pr}(\Delta^1)$ is the walking isomorphism, so we should expect that g is also a right inverse; indeed, this fact follows easily from the commutative diagram

$$\begin{array}{ccccc} \varnothing \sqcup X & \xrightarrow{! \sqcup X} & X \sqcup X & \xrightarrow{X \sqcup f} & X \sqcup Y \\ & {}_{\cong}\nwarrow & \downarrow & & \downarrow \\ & & X & = & X. \end{array}$$

4.2. Day convolution. In this section, we consider two natural tensor product operations defined on functors between symmetric monoidal categories, the left and right Day convolution tensor products. With mild assumptions on the two categories, it is well known that left convolution determines a symmetric monoidal structure on the functor category. A game of opposites then gives conditions under which right convolution determines a symmetric monoidal structure. Although these conditions are significantly more restrictive, they are nevertheless applicable in our setting of interest.

DEFINITION 4.8. Let $F, G : \mathcal{C} \to \mathcal{D}$ be functors. The left (resp. right) *Day convolution tensor product* of F and G is the left (resp. right) Kan extension $F \otimes G$ in the diagram

$$\begin{array}{ccccc} \mathcal{C} \times \mathcal{C} & \xrightarrow{F \times G} & \mathcal{D} \times \mathcal{D} & \xrightarrow{\otimes_{\mathcal{D}}} & \mathcal{D} \\ {\scriptstyle \otimes_{\mathcal{C}}}\downarrow & & & \nearrow_{F \otimes G} & \\ \mathcal{C} & & & & \end{array}$$

It should be emphasized that either convolution may fail to exist for some or all pairs of functors with fixed source and target. When the left convolution does exist, it is given explicitly by the formula

$$(F \otimes G)(C) = \operatorname{colim}\left((\otimes_{\mathcal{C}} \downarrow C) \to \mathcal{C} \times \mathcal{C} \xrightarrow{F \times G} \mathcal{D} \times \mathcal{D} \xrightarrow{\otimes_{\mathcal{D}}} \mathcal{D} \right),$$

and similarly for right convolution. Mere existence, however, does not guarantee that convolution extends to a monoidal structure.

THEOREM 4.9. *If $\mathcal{D}$ admits, and $\otimes_{\mathcal{D}}$ distributes over, colimits indexed by $(\otimes_{\mathcal{C}} \downarrow C)$ for every object $C \in \mathcal{C}$, then left Day convolution extends to a canonical symmetric monoidal structure on* $\mathrm{Fun}(\mathcal{C}, \mathcal{D})$. *In this case, there is a canonical isomorphism of categories*

$$\mathrm{Fun}^{\mathrm{lax}}(\mathcal{C}, \mathcal{D}) \cong \mathrm{Alg}_{\mathrm{Com}}(\mathrm{Fun}(\mathcal{C}, \mathcal{D})).$$

Although this result is essentially standard, it is typically stated with the unnecessarily strong assumption that $\otimes_{\mathcal{D}}$ distributes over *all* colimits. In the left handed setting, this assumption is almost always satisfied; it is in dualizing to the right handed setting in Corollary 4.12 that we require the weaker assumption. For these reasons, we have opted to include an outline of the proof—the reader may consult [**8, 17, 27**], for example, for further details.

PROOF OF THEOREM 4.9. The existence assumption grants that $F \otimes G$ is defined for every F and G. A unit is supplied by the functor $\mathbb{1}_{\mathrm{Day}}$ given by the left Kan extension of the inclusion of $\mathbb{1}_{\mathcal{D}}$ along the inclusion of $\mathbb{1}_{\mathcal{C}}$. Indeed, we claim that the left Kan extensions indicated by the dashed arrows are as claimed in the following diagrams:

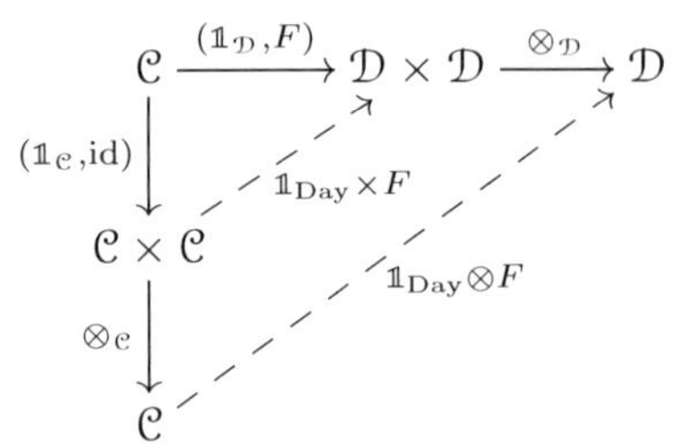

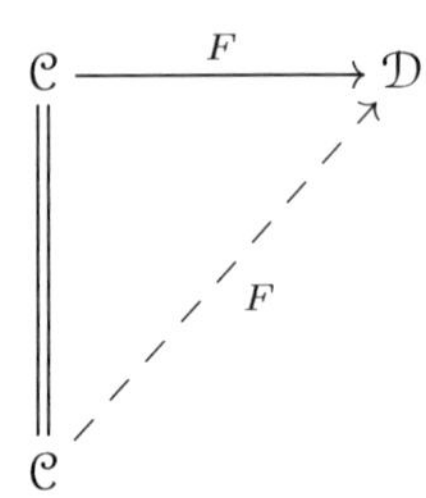

For the righthand diagram, there is nothing to show, and the identification of the innermost Kan extension in the lefthand diagram is immediate from the definitions. Invoking our assumption on distributivity of $\otimes_{\mathcal{D}}$, it follows that the outer left Kan extension is as indicated. We obtain a left unitor by observing that the left unitors of $\mathcal{C}$ and $\mathcal{D}$ give an isomorphism between the two diagrams, and similarly for a right unitor. The same considerations suffice to identify the left Kan extensions in the following diagram:

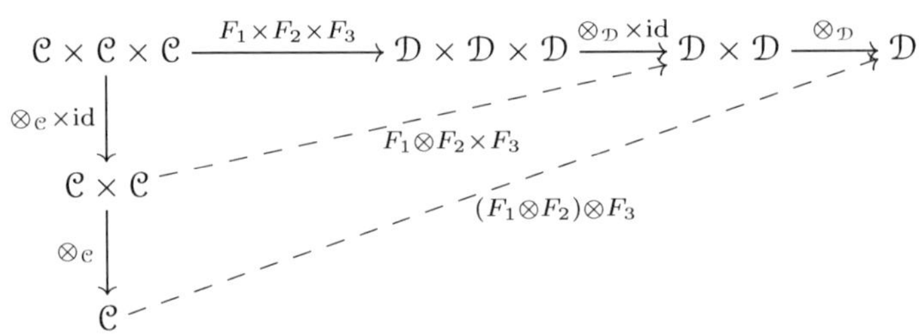

We obtain an associator after observing that the associators of $\mathcal{C}$ and $\mathcal{D}$ furnish an isomorphism between this diagram and the corresponding diagram for $F_1 \otimes (F_2 \otimes F_3)$. The same approach furnishes a symmetry. The coherence axioms for $\mathrm{Fun}(\mathcal{C}, \mathcal{D})$ follow directly from the coherence axioms for $\mathcal{C}$ and $\mathcal{D}$, since each structure morphism in the former was built from the corresponding structure morphisms in the latter two.

Toward the final claim, we observe that, by the universal property of the colimit, a monoid structure on $F : \mathcal{C} \to \mathcal{D}$ provides a compatible collection of maps $F(C_1) \otimes F(C_2) \to F(C)$, one for each arrow $C_1 \otimes C_2 \to C$. By specializing to the case $C_1 \otimes C_2 = C$, such a collection furnishes a collection of candidate components $F(C_1) \otimes F(C_2) \to F(C_1 \otimes C_2)$ of a lax structure map. Conversely, a lax structure furnishes a candidate monoid structure map via the composites

$$F(C_1) \otimes F(C_2) \to F(C_1 \otimes C_2) \to F(C).$$

Similar remarks apply to units, and it remains to verify that the associativity, symmetry, and unitality constraints of the two structures coincide, a task we leave to the reader. □

Before continuing we record a useful consequence of the proof.

COROLLARY 4.10. *If $\mathcal{D}$ admits, and $\otimes_{\mathcal{D}}$ distributes over, colimits indexed by $(\otimes_{\mathcal{C}} \downarrow C)$ for every object $C \in \mathcal{C}$, then the left Day convolution tensor product $\bigotimes_{i=1}^{n} F_i$ is canonically isomorphic to the left Kan extension in the diagram*

$$\begin{array}{ccccc} \mathcal{C}^n & \xrightarrow{(F_i)_{i=1}^n} & \mathcal{D}^n & \xrightarrow{\otimes_{\mathcal{D}}^{(n)}} & \mathcal{D} \\ \Big\downarrow {\scriptstyle \otimes_{\mathcal{C}}^{(n)}} & & & \nearrow_{\bigotimes_{i=1}^{n} F_i} & \\ \mathcal{C} & & & & \end{array}$$

The existence of the right convolution monoidal structure will rely on a technical condition on the structure of overcategories in $\mathcal{C}$, which we refer to as *sparsity*. We record several other related technical conditions for later use. In the following definition, we view finite sets as discrete categories.

DEFINITION 4.11. Let C be an object in the symmetric monoidal category $\mathcal{C}$.

(1) We say that C is *n-separable* if $(\otimes_{\mathcal{C}}^{(n)} \downarrow C)$ receives a final functor from a finite set.
(2) We say that C is *freely n-separable* if $(\otimes_{\mathcal{C}}^{(n)} \downarrow C)$ receives a Σ_n-equivariant final functor from a finite free Σ_n-set.
(3) We say that $\mathcal{C}$ is *sparse* if every object of $\mathcal{C}$ is 2-separable.

COROLLARY 4.12. *If $\mathcal{C}$ is a sparse symmetric monoidal category and $\mathcal{D}$ an additive tensor category, then right Day convolution extends to a canonical symmetric monoidal structure on* $\mathrm{Fun}(\mathcal{C}^{\mathrm{op}}, \mathcal{D})$. *In this case, there is a canonical isomorphism of categories*

$$\mathrm{Fun}^{\mathrm{oplax}}(\mathcal{C}^{\mathrm{op}}, \mathcal{D}) \cong \mathrm{Coalg}_{\mathrm{Com}}(\mathrm{Fun}(\mathcal{C}^{\mathrm{op}}, \mathcal{D})).$$

PROOF. In light of the isomorphism $\mathrm{Fun}(\mathcal{C}^{\mathrm{op}}, \mathcal{D}) \cong \mathrm{Fun}(\mathcal{C}, \mathcal{D}^{\mathrm{op}})^{\mathrm{op}}$, together with the fact that the formation of the opposite category interchanges limits and colimits, it suffices to show that $\mathcal{C}$ and $\mathcal{D}^{\mathrm{op}}$ satisfy the hypotheses of Theorem 4.9. By sparsity, a colimit indexed by $(\otimes_{\mathcal{C}} \downarrow C)$ is simply a finite coproduct in $\mathcal{D}^{\mathrm{op}}$. The proof is complete upon observing that $\mathcal{D}^{\mathrm{op}}$ is also an additive tensor category, since distributivity over finite coproducts is equivalent to distributivity over finite products by additivity. □

REMARK 4.13. The right convolution tensor product seems less well known than its sinistral cousin, perhaps due to the apparent restrictiveness of sparsity. We direct the reader to [**24**] for a prior instance of its use; one assumes that there are other antecedents, but none are known to the author.

4.3. Examples. In this section, we specialize the general theory of the preceding section to the examples of interest.

Given a collision structure S on J and a function $f : I \to J$, we write f^*S for the largest collision structure on I for which f is a map of collision structures, i.e.,

$$f^*S = \{P \in \Pi_I : f_*P \in S\}.$$

It is easy to check that f^*S, so defined, is a collision structure; note, however, that it may be empty. In the case of the inclusion of a subset $I_0 \subseteq I$, we write $S|_{I_0}$.

EXAMPLE 4.14. If $S = S_\Gamma$ is the collision structure associated to a graph Γ with vertex set I according to Example 3.5, then $S_\Gamma|_{I_0}$ is the collision structure associated to the full subgraph spanned by I_0.

EXAMPLE 4.15. If $S = S_K$ is the collision structure associated to a simplicial complex K with vertex set I according to Example 3.6, then $S_K|_{I_0}$ is the collision structure associated to the full subcomplex spanned by I_0.

Write $\mathrm{Surj}(I, n)$ for the set of surjections of I onto the set $\{1, \ldots, n\}$; in other words, $\mathrm{Surj}(I, n)$ is the set of *ordered* partitions of I with n blocks.

LEMMA 4.16. *Let S be a collision structure on I. For every $n \geq 0$, there is a canonical inclusion*

$$\mathrm{Surj}(I, n) \subseteq (\sqcup^{(n)}_{\mathcal{CSB}} \downarrow S),$$

which is final if I is non-empty.

PROOF. We define the inclusion by sending $p \in \mathrm{Surj}(I, n)$ to the map $\bigsqcup_{j=1}^n S|_{p^{-1}(j)} \to S$ of collision structures given by the canonical bijection $\bigsqcup_{j=1}^n p^{-1}(j) \cong I$. Given collision structures S_j on the finite sets I_j and a bijective map $f : \bigsqcup_{j=1}^n S_j \to S$ of collision structures, define $p_f : I \to \{1, \ldots, n\}$ by defining $p_f(i)$ to be the index j such that $i \in f(I_j)$; note that there is precisely one such index by bijectivity. Then $p_f^{-1}(j) = f(S_j)$, and we have the commutative diagram:

$$\begin{array}{ccc} \displaystyle\bigsqcup_{j=1}^{n} I_j & \xrightarrow{\ \bigsqcup_{j=1}^n f|_{I_j}\ } & \displaystyle\bigsqcup_{j=1}^{n} p_f^{-1}(j) \\ & {\scriptstyle f}\searrow \quad \swarrow{\scriptstyle \cong} & \\ & I & \end{array}$$

Finality follows upon noting that the components of the dashed arrow are uniquely determined by commutativity, each component is a map of collision structures, and no such dashed arrow exists for any $p \neq p_f$. □

COROLLARY 4.17. *If S is a collision structure on I, then S is n-separable as an object of $\mathcal{CSB}$ for every $n \geq 0$, freely if I is non-empty. In particular, $\mathcal{CSB}$ is sparse.*

PROOF. Since Σ_n acts freely on $\mathrm{Surj}(I, n)$, Lemma 4.16 directly implies that any collision structure on a non-empty finite set is freely n-separable for every $n \geq 0$. For the edge case, we note that $(\sqcup^{(n)}_{\mathcal{CSB}} \downarrow \varnothing)$ is a singleton for every $n \geq 0$. □

COROLLARY 4.18. *A graph, a simplicial complex, or a finite set is n-separable as an object of $\mathcal{GB}$, $\mathcal{SCB}$, or $\mathcal{FB}$, respectively, for every $n \geq 0$, freely if non-empty. In particular, all three categories are sparse.*

PROOF. First, the inclusion of $\mathrm{Surj}(I, n)$ of Lemma 4.16 factors through $(\sqcup^{(n)}_{\mathcal{GB}} \downarrow \Gamma)$, $(\sqcup^{(n)}_{\mathcal{SCB}} \downarrow K)$, or $(\sqcup^{(n)}_{\mathcal{FB}} \downarrow I)$, respectively. For the first two, this claim follows from Examples 4.14 and 4.15, and for the third it is obvious. Since these three subcategories are full, each inclusion is also final. □

Combining these corollaries with Corollary 4.12, we see that it is sensible to contemplate right Day convolution in this context. Corollary 4.10 and Lemma 4.16 yield the following formula for this tensor product.

COROLLARY 4.19. *Let $\mathcal{D}$ be an additive tensor category. Given functors $F_j : \mathcal{CSB}^{\mathrm{op}} \to \mathcal{D}$ for $1 \leq j \leq n$ and a collision structure S on I, there is a canonical natural isomorphism*

$$\left(\bigotimes_{j=1}^{n} F\right)(S) \cong \bigoplus_{I=\bigsqcup_{j=1}^{n} I_j} \bigotimes_{j=1}^{n} F(S|_{I_j}).$$

Examining the proof of Corollary 4.18, we see that the same formula holds for $\mathcal{GB}$ and $\mathcal{SCB}$ (and $\mathcal{FB}$, of course). By Examples 4.14 and 4.15, this formula may be interpreted geometrically in terms of full subgraphs and full subcomplexes, respectively.

5. Twisted Koszul duality

The purpose of this section is to explain how the theory of Koszul duality between Lie algebras and cocommutative coalgebras extends to the twisted setting—see Theorem 5.15. Classically, this theory is developed under two key assumptions: first, that the ground ring is a field of characteristic zero; second, that the graded objects in question are sufficiently connected [**12, 33**]. These assumptions serve to guarantee the good behavior of divided powers—their homotopy invariance and nilpotence, respectively. In our setting, the same good behavior is guaranteed instead by the presence of a *weighting* with certain natural properties—see Section 5.1. Otherwise, the arguments are essentially identical, and we will at times allow ourselves a certain brevity.

We work throughout over a fixed ground field $\mathbb{F}$, a limitation easily overcome by judicious assumptions of flatness.

5.1. Weights. In this section, we introduce a piece of extra structure that will serve as a substitute for connectivity and divisibility. In the next definition, we refer to Definition 4.11.

DEFINITION 5.1. Let $\mathcal{C}$ be a monoidal category.

(1) A *weighting* of $\mathcal{C}$ is a lax monoidal functor $w : \mathcal{C} \to \mathbb{Z}_{\geq 0}$. The value $w(C)$ is the *weight* of C. The pair $(\mathcal{C}, w)$ is a *weighted monoidal category*
(2) A *combinatorial monoidal category* is a weighted monoidal category such that $C \in \mathcal{C}$ is freely n-separable for every $n \geq 0$ whenever $w(C) > 0$.

Concretely, a weighting amounts to the assignment of a non-negative integer $w(C)$ to each $C \in \mathcal{C}$ such that the following inequalities hold for every $C_1, C_2 \in \mathcal{C}$:

(1) $w(C_1) \leq w(C_2)$ whenever $\mathrm{Hom}_{\mathcal{C}}(C_1, C_2) \neq \varnothing$
(2) $w(C_1) + w(C_2) \leq w(C_1 \otimes C_2)$.

EXAMPLE 5.2. Recording the cardinality of a finite set determines a weighting of $\mathcal{FB}$, whence of $\mathcal{CSB}$, $\mathcal{GB}$, and $\mathcal{SCB}$. By the results of Section 4.3, these weighted monoidal categories are combinatorial.

DEFINITION 5.3. Let $\mathcal{C}$ be a weighted monoidal category and $\mathcal{D}$ an additive tensor category. We say that a functor $F : \mathcal{C} \to \mathcal{D}$ is *concentrated in weight* $\geq k$

if $F(C) = 0$ whenever $w(C) < k$. If F is concentrated in weight ≥ 1, then we say that F is *reduced.*

This definition is justified by the following simple result.

LEMMA 5.4. *If $\mathcal{D}$ is an additive tensor category, then $\mathcal{D}$ admits a zero object 0, and $0 \otimes D \cong 0$ for every $D \in \mathcal{D}$.*

PROOF. Additivity guarantees the existence of the biproduct indexed by the empty set, which is to say a zero object. Second, since the tensor product of $\mathcal{D}$ distributes over finite coproducts, we have

$$\begin{aligned} 0 \otimes D &\cong \operatorname{colim}\left(\varnothing \xrightarrow{!} \mathcal{D}\right) \otimes D \cong \operatorname{colim}\left(\varnothing \times \mathcal{D} \xrightarrow{! \times \mathrm{id}} \mathcal{D} \times \mathcal{D} \xrightarrow{\otimes} \mathcal{D}\right) \\ &\cong \operatorname{colim}\left(\varnothing \xrightarrow{!} \mathcal{D}\right) \cong 0. \end{aligned}$$

□

The comparison to connectivity made above is justified by the following result.

LEMMA 5.5. *Let $\mathcal{C}$ be a sparse weighted symmetric monoidal category. If $F_i : \mathcal{C}^{\mathrm{op}} \to \mathcal{D}$ is concentrated in weight $\geq k_i$ for $i \in \{1, 2\}$, then $F_1 \otimes F_2$ is concentrated in weight $\geq k_1 + k_2$.*

PROOF. Let $C \in \mathcal{C}$ be an object with $w(C) < k_1 + k_2$. From our assumption of sparsity and the definition of right Day convolution, there is a collection of arrows $C_1 \otimes C_2 \to C$ such that

$$(F \otimes G)(C) \cong \bigoplus_{C_1 \otimes C_2 \to C} F_1(C_1) \otimes F_2(C_2) = 0,$$

so it suffices to show that each summand vanishes. We have the inequality

$$k_1 + k_2 > w(C) \geq w(C_1 \otimes C_2) \geq w(C_1) + w(C_2).$$

Since weights are non-negative, it follows that $w(C_1) < k_1$ or $w(C_2) < k_2$, whence $F_1(C_1) = 0$ or $F_2(C_2) = 0$. □

The comparison to divisibility made above is justified by the following result.

LEMMA 5.6. *Let $\mathcal{C}$ be a combinatorial symmetric monoidal category and $\tau : F_1 \to F_2$ a natural transformation between reduced functors from $\mathcal{C}^{\mathrm{op}}$ to the category of chain complexes in an Abelian tensor category. If τ is a quasi-isomorphism in weight $< n$, then $\Gamma^k(\tau)$ is a quasi-isomorphism in weight $< n + k - 1$ for every $k > 0$. In particular, if $k > 1$, then $\Gamma^k(\tau)$ is a quasi-isomorphism in weight $\leq n$.*

PROOF. Fix $C \in \mathcal{C}$ with $w(C) < n + k - 1$. If $w(C) = 0$, then $T^k(F_i)(C) = 0$ by Lemma 5.5, and there is nothing to show, so assume otherwise. Then C is freely n-separable by our assumption on $\mathcal{C}$, so Corollary 4.10 grants the existence of a finite, Σ_k-free set $\{f_r : C_{1,r} \otimes \cdots \otimes C_{k,r} \to C\}$ of morphisms such that

$$T^k(F_i)(C) \cong \bigoplus_r \bigotimes_{j=1}^k F_i(C_{j,r})$$

compatibly with $T^k(\tau)$. Since the F_i are reduced, we may assume without loss of generality that $w(C_{j,r}) > 0$ for every j and r. It then follows from the inequality

$$n+k-1 > w(C) \geq \sum_{j=1}^{k} w(C_{j,r})$$

that $w(C_{j,r}) < n$ for every j and r. By assumption, $\tau_{C_{j,r}}$ is a quasi-isomorphism, so $T^k(F_i)(C)$ is a quasi-isomorphism. By freeness, this quasi-isomorphism descends to a quasi-isomorphism on Σ_k-invariants, completing the proof. □

REMARK 5.7. In a previous version of this paper, Lemma 5.6 was stated with the weaker bound only. We thank the referee for pointing out that the proof establishes the improved bound.

5.2. Twisted (co)algebraic structures. We come now to the main definitions.

DEFINITION 5.8. Let $\mathcal{C}$ be a sparse symmetric monoidal category. A *$\mathcal{C}$-twisted Lie algebra* (over $\mathbb{F}$) is a Lie algebra in $\mathrm{Fun}(\mathcal{C}^{\mathrm{op}}, \mathcal{C}\mathrm{h}_{\mathbb{F}})$, regarded as symmetric monoidal under right Day convolution.

Similarly, one has the notion of a $\mathcal{C}$-twisted cocommutative coalgebra, and so on. In the case $\mathcal{C} = \mathcal{FB}$, one recovers the classical notion of a twisted (co)algebraic structure.

REMARK 5.9. As the reader may know, the term "Lie algebra" has several inequivalent definitions. Ultimately, we will specialize to a setting in which all such definitions coincide, so the reader is welcome to imagine that her favorite definition is also ours.

Given a $\mathcal{C}$-twisted cocommutative coalgebra K, we write $\overline{K}$ for the kernel of the counit $K \to \mathbb{1}$.

LEMMA 5.10. *Let L be a $\mathcal{C}$-twisted Lie algebra and K a $\mathcal{C}$-twisted cocommutative coalgebra.*

(1) *The unique degree -1 coderivation ∂_{CE} of $\Gamma(L[1])$ extending the composite*

$$\Gamma(L[1])[-1] \to \Gamma^2(L[1])[-1] \xrightarrow{[-,-][1]} L[1]$$

squares to zero.

(2) *The unique degree -1 derivation ∂_{Q} of $\mathrm{Lie}(\overline{K}[-1])$ extending the composite*

$$\overline{K}[-1] \xrightarrow{\Delta[-1]} \mathrm{Lie}^2(\overline{K}[-1])[1] \subseteq \mathrm{Lie}(\overline{K}[-1])[1]$$

squares to zero.

PROOF. We give only a brief outline, the proof being entirely parallel to the corresponding portions of [**12**, IV.22(b),(e)]. For the first claim, since the square of a coderivation of odd degree is again a coderivation, and since $\Gamma(L[1])$ is cofree (as a conilpotent cocommutative coalgebra), it suffices to verify that the composite

$$\Gamma^3(L[1])[-2] \to \Gamma^2(L[1])[-1] \to L[1]$$

vanishes, which is equivalent to the Jacobi identity. The proof of the second claim is essentially the same save that we appeal instead to coassociativity. □

This result permits the following fundamental definition.

Definition 5.11. Let L be a $\mathcal{C}$-twisted Lie algebra and K a $\mathcal{C}$-twisted cocommutative coalgebra.

(1) The *Chevalley–Eilenberg complex* of L is the $\mathcal{C}$-twisted graded cocommutative coalgebra
$$\mathrm{CE}(L) = (\Gamma(L[1]), \partial_{\mathrm{CE}} + \partial_L).$$
Its homology is called the *Lie algebra homology* of L and denoted $H_*^{\mathrm{Lie}}(L)$.

(2) The *Quillen complex* of K is the $\mathcal{C}$-twisted graded Lie algebra
$$\mathrm{Q}(K) = (\mathrm{Lie}(\overline{K}[-1]), \partial_{\mathrm{Q}} + \partial_K).$$

Remark 5.12. Some readers may be surprised at the appearance of divided powers rather than symmetric powers in the Chevalley–Eilenberg complex. The two coincide in characteristic zero, and more generally in any setting in which the norm map is an isomorphism on tensor powers (as it will be below).

These two complexes are closely interrelated.

Lemma 5.13. *Let L be a $\mathcal{C}$-twisted Lie algebra and K a $\mathcal{C}$-twisted cocommutative coalgebra.*

(1) *The map $\eta : K \to \mathrm{CE}(\mathrm{Q}(K))$ of $\mathcal{C}$-twisted coalgebras induced by the inclusion $\iota : \overline{K} \cong \overline{K}[-1][1] \subseteq \mathrm{Q}(K)[1]$ is a chain map.*

(2) *The map $\epsilon : \mathrm{Q}(\mathrm{CE}(L)) \to L$ of $\mathcal{C}$-twisted Lie algebras induced by the projection $\pi : \mathrm{CE}(L)[-1] \to L[1][-1] \cong L$ is a chain map.*

Proof. Again, we give only a brief outline, as the details are parallel to the relevant portions of the proof of [**12**, Thm. 22.9]. Writing δ_1 and δ_2, respectively, for the clockwise and counterclockwise composites in the diagram
$$\begin{array}{ccc} \overline{K} & \xrightarrow{\eta} & \mathrm{CE}(\mathrm{Q}(K)) \\ {\scriptstyle \partial_K}\downarrow & & \downarrow{\scriptstyle \partial_{\mathrm{CE}}+\partial_{\mathrm{Q}(K)}} \\ \overline{K} & \xrightarrow{\eta} & \mathrm{CE}(\mathrm{Q}(K)), \end{array}$$
the first claim is the equality $\delta_1 = \delta_2$. Since η is a map of coalgebras, the δ_i are coderivations; therefore, since $\mathrm{CE}(\mathrm{Q}(K))$ is cofree, it suffices to check that the composite
$$\overline{K} \xrightarrow{\delta_i} \mathrm{CE}(\mathrm{Q}(K)) \xrightarrow{\pi} \mathrm{Q}(K)[1]$$
is independent of i, which is essentially a tautology. The proof of the second claim is similar. □

5.3. Duality. In order to state the main result, we require one further definition.

Definition 5.14. Let $\mathcal{C}$ be a sparse weighted symmetric monoidal category. We say that a $\mathcal{C}$-twisted Lie algebra is *reduced* if its underlying functor is so. We say that a $\mathcal{C}$-twisted cocommutative coalgebra K is *reduced* if $\overline{K}$ is so.

The goal of this section is to prove the following analogue of [**12**, Thm. 22.9].

Theorem 5.15. *If $\mathcal{C}$ is combinatorial, then the maps η and ϵ are quasi-isomorphisms on reduced objects.*

The main ingredients are the following two lemmas.

LEMMA 5.16. *If $\mathcal{C}$ is combinatorial, then the Chevalley–Eilenberg complex preserves and reflects quasi-isomorphisms between reduced $\mathcal{C}$-twisted Lie algebras.*

PROOF. Let $\varphi : L_1 \to L_2$ be a map of $\mathcal{C}$-twisted Lie algebras. Assuming that φ is a quasi-isomorphism, Lemma 5.6 grants that $\Gamma^k(\varphi[1])$ is also a quasi-isomorphism for every $k \geq 0$. Filtering the Chevalley–Eilenberg complex by tensor degree, we conclude that φ induces a quasi-isomorphism at the level of associated graded complexes. By induction and the five lemma, it follows that $\mathrm{CE}(\varphi)$ is a quasi-isomorphism after restriction to any filtration stage; therefore, since direct limits preserve quasi-isomorphisms, $\mathrm{CE}(\varphi)$ itself is a quasi-isomorphism.

For reflection, we adapt the argument of [**14**, 4.1.9]. Assuming that $\mathrm{CE}(\varphi)$ is a quasi-isomorphism, let $C \in \mathcal{C}$ be of minimal weight such that φ_C is not a quasi-isomorphism. Since the L_i are reduced, we have $n := w(C) > 0$. Setting $R_i := \mathrm{coker}(L_i[1] \to \mathrm{CE}(L_i))$, we have the following commutative diagram of chain complexes with exact rows:

$$\begin{array}{ccccccccc} 0 & \longrightarrow & L_1(C)[1] & \longrightarrow & \mathrm{CE}(L_1)(C) & \longrightarrow & R_1(C) & \longrightarrow & 0 \\ & & \downarrow{\scriptstyle \varphi_C[1]} & & \downarrow{\scriptstyle \mathrm{CE}(\varphi)_C} & & \downarrow & & \\ 0 & \longrightarrow & L_2(C)[1] & \longrightarrow & \mathrm{CE}(L_2)(C) & \longrightarrow & R_2(C) & \longrightarrow & 0. \end{array}$$

It suffices to show that the righthand vertical arrow is a quasi-isomorphism; indeed, since the middle arrow is a quasi-isomorphism by assumption, the five lemma then implies that the lefthand arrow is a quasi-isomorphism, a contradiction. To this end, we filter R_i by tensor degree, as above, and observe that the associated graded pieces are of the form $\Gamma^k(L_i[1])$ for $k > 1$. By minimality, φ is a quasi-isomorphism in weight $< n$, so Lemma 5.6 implies that $\Gamma^k(L_i[1])$ is a quasi-isomorphism in weight $\leq n$ for $k > 1$. In particular, the righthand vertical arrow above induces a quasi-isomorphism at the level of associated graded complexes, and the same argument as before completes the proof. □

LEMMA 5.17. *The projection $\rho : \mathrm{CE}(\mathrm{Lie}(V)) \to V[1] \oplus \mathbb{1}$ is a quasi-isomorphism.*

PROOF. Since adjunctions compose, the universal enveloping algebra of $\mathrm{Lie}(V)$ is the tensor algebra $T(V)$; therefore, the homology of the Chevalley–Eilenberg complex is canonically isomorphic to $\mathrm{Tor}_*^{T(V)}(\mathbb{1}, \mathbb{1})$, which may be computed from the two-step complex $T(V) \otimes V \to T(V)$. The claim follows immediately. □

With these results in hand, we complete the proof.

PROOF OF THEOREM 5.15. Filtering $\mathrm{Q}(K)$ by bracket length, the associated graded Lie algebra is $\mathrm{Lie}(\overline{K}[-1])$, and the induced filtration on $\mathrm{CE}(\mathrm{Q}(K))$ has associated graded object $\mathrm{CE}(\mathrm{Lie}(K[-1]))$. The composite

$$\overline{K} \xrightarrow{\mathrm{gr}(\eta)} \mathrm{CE}(\mathrm{Lie}(\overline{K}[-1])) \xrightarrow{\rho} \overline{K}$$

is the identity, and ρ is a quasi-isomorphism by Lemma 5.17, so $\mathrm{gr}(\eta)$ is also a quasi-isomorphism. It follows as before that η is a quasi-isomorphism. It is easy to check that $\mathrm{CE}(\epsilon) \circ \eta = \mathrm{id}$, so it follows that $\mathrm{CE}(\epsilon)$ is a quasi-isomorphism, and Lemma 5.16 implies that ϵ is also a quasi-isomorphism. □

REMARK 5.18. In classical operadic Koszul duality [**26**], the cobar construction (playing the role of the Quillen complex) fails to preserve quasi-isomorphisms in general, so one obtains instead an equivalence between standard homotopy theory on the algebra side and a nonstandard homotopy theory on the coalgebra side. From this point of view, the main point of this section is that Q *does* preserve quasi-isomorphisms in our setting, which is the essential content of the reflection clause of Lemma 5.16.

6. Projection spaces

We arrive at last at our primary object of study. After cataloguing examples of projection spaces of interest, we define the functor of rational primitives and prove Theorems 1.2 and 1.3 as stated in the introduction.

6.1. Examples. We recall that, according to Definition 1.1, a projection space (over $\mathcal{C}$) is a topological presheaf on $\mathrm{Pr}(\mathcal{C})$. We begin with the most basic example.

EXAMPLE 6.1. Viewing a finite set as a discrete topological space defines a functor $\mathcal{FI} \to \mathcal{T}\mathrm{op}$. By restriction along this functor, any presheaf of spaces on $\mathcal{T}\mathrm{op}$ determines a projection space over $\mathcal{FB}$ (recall Proposition 2.8); in particular, this construction applies to the representable presheaf $\mathrm{Map}(-,X)$ for any topological space X. Under the identification $\mathrm{Map}(I,X)\cong X^I$, one checks that the structure map associated to the injection $f: I\to J$ is the projection $\pi_f : X^J \to X^I$, whose ith component is the projection onto the factor of X^J indexed by $f(i)$.

Although this example is rather uninteresting, it gives rise to the main motivating example.

EXAMPLE 6.2. Given a finite set, the configuration space of I-indexed points in X is defined as the subspace $\mathrm{Conf}_I(X)\subseteq X^I$ of injective functions $I\to X$. Since the restriction of an injection along an injection is again an injection, the dashed filler exists in the diagram

$$\begin{array}{ccc} \mathrm{Conf}_J(X) & \longrightarrow & X^J \\ \downarrow & & \downarrow{\scriptstyle \pi_f} \\ \mathrm{Conf}_I(X) & \longrightarrow & X^I, \end{array}$$

so the collection of all such configuration spaces inherits the structure of a projection space over $\mathcal{FB}$.

The next two examples generalize the previous two in the presence of an action by a group G.

EXAMPLE 6.3. Applying the considerations of Example 6.1 to the inclusion $\mathcal{FI}_G \to \mathcal{T}\mathrm{op}_G$ and the functor $\mathrm{Map}_G(-,X)$ represented by the G-space X, we obtain a projection space over $\mathcal{FB}_G$ (recall Corollary 2.10). Concretely, the value of this projection space on the finite free G-set I is X^{I_0}, where we have written $I\cong G\times I_0$ non-canonically. The structure maps combine the action of G with the projections of Example 6.1.

EXAMPLE 6.4. Given a finite free G-set I, the orbit configuration space of I-indexed points in X is defined as the subspace $\mathrm{Conf}_I^G(X)\subseteq \mathrm{Map}_G(I,X)$ of injective equivariant functions. Equivalently, writing $I\cong G\times I_0$ (non-canonically), the orbit configuration space is the subspace of X^{I_0} consisting of tuples with pairwise disjoint orbits, i.e., it is the pullback

$$\begin{array}{ccc} \mathrm{Conf}_I^G(X) & \longrightarrow & X^{I_0} \\ \big\downarrow & & \big\downarrow \\ \mathrm{Conf}_{I_0}(X_G) & \longrightarrow & (X_G)^{I_0}. \end{array}$$

Since equivariant injections are closed under composition, the first description endows the collection of orbit configuration spaces with the structure of a projection space over $\mathcal{FB}_G$.

There is also a linear analogue of these examples. We fix a topological field $\mathbb{F}$, the cases of $\mathbb{R}$ and $\mathbb{C}$ being of greatest interest.

EXAMPLE 6.5. By restricting along the inclusion of $\mathcal{VI}_{\mathbb{F}}$ into the category of topological vector spaces, a presheaf of spaces on the latter determines a projection space over $\mathcal{VB}_{\mathbb{F}}$ (recall Corollary 2.10); in particular this construction applies to the representable presheaf $\mathrm{Hom}_{\mathbb{F}}(-, W)$ for any topological vector space W. Since the restriction of an injection along an injection is again an injection, the dashed filler exists in the diagram

$$\begin{array}{ccc} V_n(W) & \longrightarrow & \mathrm{Hom}_{\mathbb{F}}(\mathbb{F}^n, W) \\ \big\downarrow & & \big\downarrow {\scriptstyle \varphi^\vee} \\ V_m(W) & \longrightarrow & \mathrm{Hom}_{\mathbb{F}}(\mathbb{F}^m, W), \end{array}$$

where $V_n(W)$ denotes the space of linear injections of $\mathbb{F}^n$ into W, which is simply the usual non-compact Stiefel manifold of n-frames in the real or complex case.

REMARK 6.6. One imagines that our framework expands easily to encompass algebrogeometric examples such as flag varieties in positive characteristic.

Given a partition P of I, we write $\Delta_P = \{(x_i)_{i \in I} : i_1 \sim_P i_2 \implies x_{i_1} = x_{i_2}\}$. Given a collision structure S on I, we set $\Delta_S = \bigcup_{P \in S} \Delta_P$ and define the generalized configuration space [**32**] associated to S as

$$\mathrm{Conf}_S(X) = X^I \setminus \Delta_S.$$

Notice that, if $P \leq P'$, then $\Delta_{P'} \subseteq \Delta_P$, so $\Delta_S = \bigcup_{P \in S_0} \Delta_P$ for any generating set $S_0 \subseteq S$.

LEMMA 6.7. *Let S and T be collision structures on I and J, respectively, and $f : I \to J$ an injective map of collision structures. The dashed filler exists in the diagram*

$$\begin{array}{ccc} \mathrm{Conf}_T(X) & \longrightarrow & X^J \\ \big\downarrow & & \big\downarrow {\scriptstyle \pi_f} \\ \mathrm{Conf}_S(X) & \longrightarrow & X^I. \end{array}$$

PROOF. By our assumption on f, we have the containment $\Delta_{f_*S} \subseteq \Delta_T$. On the other hand, we calculate that

$$\begin{aligned}
\pi_f^{-1}(\Delta_S) &= \bigcup_{P\in S} \pi_f^{-1}(\Delta_P) \\
&= \bigcup_{P\in S} \{(x_j)_{j\in J} : i_1 \sim_P i_2 \implies x_{f(i_1)} = x_{f(i_2)}\} \\
&= \bigcup_{P\in S} \{(x_j)_{j\in J} : f(i_1) \sim_P f(i_2) \implies x_{f(i_1)} = x_{f(i_2)}\} \\
&= \bigcup_{P\in S} \{(x_j)_{j\in J} : j_1 \sim_{f_*P} j_2 \implies x_{j_1} = x_{j_2}\} \\
&= \bigcup_{P\in S} \Delta_{f_*P} \\
&= \Delta_{f_*S},
\end{aligned}$$

where the last equality uses that $\{f_*P : P \in S\}$ is a generating set for f_*S by definition. Thus, we have the containment $\mathrm{Conf}_T(X) \subseteq X^J \setminus \pi_f^{-1}(\Delta_S) = \pi_f^{-1}(\mathrm{Conf}_S(X))$, as desired. □

EXAMPLE 6.8. Pulling the projection space $I \mapsto X^I$ of Example 6.1 back along the forgetful functor from collision structures to finite sets, we obtain a projection space over $\mathcal{CSB}$ with the same values. Invoking Lemma 6.7, we obtain a projection space over $\mathcal{CSB}$ extending the assignment $S \mapsto \mathrm{Conf}_S(X)$ on objects.

REMARK 6.9. There is an obvious common generalization of Examples 6.4 and 6.8, which one would likely dub a generalized orbit configuration space. The ensemble of such also forms a projection space, where one would likely dub the relevant combinatorial object a G-collision structure.

EXAMPLE 6.10. Restricting the projection space of Example 6.8 to $\mathcal{GJ}$, we obtain a projection space over $\mathcal{GB}$ extending the assignment $\Gamma \mapsto \mathrm{Conf}_{S_\Gamma}(X)$ on objects.

EXAMPLE 6.11. Restricting the projection space of Example 6.8 to $\mathcal{SCJ}$, we obtain a projection space over $\mathcal{SCB}$ extending the assignment $K \mapsto \mathrm{Conf}_{S_K}(X)$ on objects.

As the reader will easily verify, all of the examples of projection spaces given in this section are reduced.

6.2. Proofs of the main results. In the following definition, the reader may take the functor A_* to be any symmetric replacement for the oplax monoidal functor rational singular chains. A specific example of such a replacement is given below in Appendix A.

DEFINITION 6.12. Let $\mathcal{C}$ be a sparse symmetric monoidal category and X a projection space over $\mathcal{C}$. The $\mathcal{C}$-twisted Lie algebra of (derived) *rational primitives* of X is the value on X of the composite functor

$$\begin{array}{l} L : \mathrm{Fun}(\mathrm{Pr}(\mathcal{C})^{\mathrm{op}}, \mathcal{T}\mathrm{op}) \xrightarrow{(4.3)} \mathrm{Fun}^{\mathrm{oplax}}(\mathcal{C}^{\mathrm{op}}, \mathcal{T}\mathrm{op}) \\ \qquad \downarrow A_* \\ \mathrm{Fun}^{\mathrm{oplax}}(\mathcal{C}^{\mathrm{op}}, \mathcal{C}\mathrm{h}_{\mathbb{Q}}) \xrightarrow{(4.12)} \mathrm{Coalg}_{\mathrm{Com}}(\mathrm{Fun}(\mathcal{C}^{\mathrm{op}}, \mathcal{C}\mathrm{h}_{\mathbb{Q}})) \\ \qquad \downarrow \mathrm{Q} \\ \mathrm{Alg}_{\mathrm{Lie}}(\mathrm{Fun}(\mathcal{C}^{\mathrm{op}}, \mathcal{C}\mathrm{h}_{\mathbb{Q}})). \end{array}$$

PROOF OF THEOREM 1.2. Since the isomorphisms of Corollaries 4.3 and 4.12 do not change the underlying functor, we have

$$\begin{aligned} H_*^{\mathrm{Lie}}(L(X)) &= H_*(\mathrm{CE}(\mathrm{Q}(A_*(X)))) \\ &\cong H_*(A_*(X)) \\ &\cong H_*(X; \mathbb{Q}) \end{aligned}$$

by Theorem 5.15 and Corollary A.5. □

We turn now to stability. The prevailing philosophy that has emerged in the wake of the discovery of representation stability is that stability phenomena are the concrete consequences of finite generation results, representation stability itself being equivalent to finite generation over $\mathcal{FI}$ [**6**]. Prerequisitely, then, one requires an action.

A feature of the Chevalley–Eilenberg complex is that it is a symmetric monoidal functor, converting products into tensor products. Thus, as explained at greater length in [**25**], a central Lie subalgebra $L_0 \subseteq L$ gives rise to an action of $\Gamma(L_0[1])$ on Lie algebra homology, while an Abelian quotient $L \to L_1$ gives rise to an action of $\mathrm{Sym}(L_1[1])$ on Lie algebra cohomology. Through Theorem 1.2, these two constructions give rise to a host of actions on the (co)homology of a projection space, each a potential source of stability phenomena.

REMARK 6.13. In the case $\mathcal{C} = \mathbb{Z}_{\geq 0}$, as shown in [**25**], these two types of action are united in the action of a single algebra, called a *transit algebra*. It is less clear how to describe their interaction in general, especially when $\mathcal{C}$ is not a groupoid.

We now show that this potential is often realized in the classical case $\mathcal{C} = \mathcal{FB}$. A functor from this category (or its opposite) is a symmetric sequence, i.e., a list of objects indexed by the non-negative integers (their weights), together with a Σ_k-action on the kth object for each $k \geq 0$. We indicate the weight with a subscript, and we write $V(1)$ for the symmetric sequence with $V(1)_1 = V$ and $V(1)_k = 0$ for $k \neq 1$.

THEOREM 6.14. *If X is a reduced $\mathcal{FI}^{\mathrm{op}}$-space, then $H^*(X)$ is canonically a $\mathrm{Sym}(H^0(X_1)(1))$-module. This module is finitely generated provided $L(X)$ satisfies the following conditions.*

(1) *$H_i(L(X)_k)$ is finite dimensional for every $i \geq 0$ and $k \geq 0$.*
(2) *$H_{-1}(L(X)_k) = 0$ for $k > 1$ and $H_i(L(X)) = 0$ for $i < -1$.*
(3) *$H_i(L(X)_k) = 0$ for i fixed and k sufficiently large.*

It is well known that representation stability is equivalent to finite generation over the free twisted commutative algebra on a single generator in degree 0 and weight 1, so Theorem 1.3 is a special case of this result. In general, the finite generation of the theorem implies a kind of generalized representation stability [**34**].

REMARK 6.15. We briefly contextualize the assumptions of the theorem. The third assumption is that $L(X)$ is *eventually highly connected*; one should view this assumption, which resembles standard hypotheses in the study of homological stability, as the key assumption. The first assumption is simply that each component space is of finite type, as will be the case in most examples of interest. The second assumption, which may at first appear the strangest, is in fact also quite reasonable; indeed, it is not hard to show that it holds whenever X is path connected in each weight, or more generally when $H_0(X)$ is cogenerated cofreely by $H_0(X_1)$, as is the case for configuration spaces. One could weaken this assumption at the cost of a more involved statement.

LEMMA 6.16. *If X is a reduced $\mathcal{FJ}^{\mathrm{op}}$-space, then $L(X)$ admits the Abelian twisted Lie algebra $H_0(X_1)(1)[-1]$ as a canonical quotient.*

PROOF. We begin by observing the canonical isomorphism $L(X)_1 \cong A_*(X_1)[-1]$ of chain complexes. Extending by zero, we obtain the composite map

$$L(X) \to A_*(X_1)(1)[-1] \to H_0(X_1)(1)[-1],$$

where the second composition factor is the projection to the quotient. This map is clearly surjective, and it is a map of Lie algebras; indeed, $L(X)$ is reduced, and weight is additive under the bracket. $\square$

PROOF OF THEOREM 6.14. The Chevalley–Eilenberg complex is the total complex of a bicomplex, whose differentials reflect the internal differential of the Lie algebra and the Lie bracket, respectively. One of the two spectral sequences associated to the dual of this bicomplex has the form

$$E_1 \cong \mathrm{Sym}(H^*(L(X))[1]) \implies H^*(X;\mathbb{Q})$$

(we use our first assumption to guarantee that the dual of the divided power is the symmetric power). This spectral sequence is a spectral sequence of $\mathrm{Sym}(H^0(X_1)(1))$-modules by functoriality, since the action arises from a map of Lie algebras. As such a module, the E^1-page is freely generated by $\mathrm{Sym}(V)$, where

$$V = \frac{H^*(L(X))[1]}{H^0(L(X_1)(1))[1]}.$$

By our first assumption, $H^0(X_1)$ is finite dimensional, so $\mathrm{Sym}(H^0(X_1)(1))$ is Noetherian [**37**]; thus, it suffices to show that $\mathrm{Sym}(V)$ is finite dimensional in fixed degree i, for which we calculate that

$$\mathrm{Sym}(V)_{k,i} = \bigoplus_{r\geq 0}\left(\bigoplus_{k_1+\cdots+k_r=k}\ \bigoplus_{i_1+\cdots+i_r=i} \mathrm{Ind}_{\prod_{j=1}^r \Sigma_{k_j}}^{\Sigma_k} \bigotimes_{j=1}^{r} V_{k_j,i_j}\right)_{\Sigma_r}.$$

For fixed k, this expression is finite dimensional by reducedness and our first assumption, so it suffices to show that it vanishes for k sufficiently large. By our third assumption on $L(X)$, there exists ℓ sufficiently large so that $V_{k,\leq i} = 0$ for $k \geq \ell$; therefore, the summand indexed by r vanishes for $k \geq r\ell$. On the other hand, our second assumption on $L(X)$ implies that $V_{k,i} = 0$ for $i \leq 0$, so we may take $i_j > 0$ for $1 \leq j \leq r$ in the above expression. It follows that the summands indexed by $r > i$ all vanish. Combining these observations, we conclude that $\mathrm{Sym}(V)_{k,i} = 0$ for $k \geq i\ell$, as desired. □

REMARK 6.17. Let M be a (for simplicity) orientable manifold of dimension n. According to the conjecture articulated in Section 1.3(2), the homology of the Lie algebra of rational primitives of the ordinary configuration spaces of M can be described in weight k as the vector space $H^{-*}(M) \otimes \mathrm{Lie}(k)[k(n-1)]$, where $\mathrm{Lie}(k)$ is the kth Lie representation. As long as $n > 1$, this Lie algebra satisfies the assumptions of Theorem 6.14, so we recover the primordial example of representation stability. This analysis does not actually depend on the validity of the conjecture, since the Chevalley–Eilenberg complex of this Lie algebra does calculate the correct homology.

Appendix A. Sullivan chains

The cup product arises from the Alexander–Whitney map $C_*(X \times Y) \to C_*(X) \otimes C_*(Y)$, a natural transformation constituting an oplax monoidal structure that is famously not symmetric. Fortunately, the failure of symmetry is governed by highly coherent homotopies, which are strictifiable in characteristic zero for abstract reasons.

The purpose of this appendix is to present an explicit model for this strictification. In brief, we construct a pre-dual of (a completion of) the functor of Sullivan cochains A^*, a reference for which is [**12**, II.10(c)], where it is denoted A_{PL}. It is common to discuss this complex in barycentric coordinates, but the expression in increasing coordinates given in [**36**] will be more convenient for our purposes.

Write V_n for the chain complex with basis $\{x_1, \ldots, x_n, dx_1, \ldots, dx_n\}$, where $|x_i| = 0$. This complex carries a natural simplicial structure via the maps

$$\partial_i(x_k) = \begin{cases} x_k & k \leq i \\ x_{k-1} & k > i \end{cases} \qquad s_j(x_k) = \begin{cases} x_k & k \leq j \\ x_{k+1} & k > j. \end{cases}$$

The dual $V_n^\vee$ thus carries a natural cosimplicial structure.

DEFINITION A.1. Fix $n \geq 0$.

(1) The complex of *Sullivan cochains* on the standard n-simplex Δ^n is the commutative differential graded algebra $A^*(\Delta^n)$ freely generated by V_n.
(2) The complex of *Sullivan chains* on the standard n-simplex Δ^n is the conilpotent cocommutative differential graded coalgebra cofreely cogenerated by $V_n^\vee$.
(3) The complex of *completed Sullivan cochains* on Δ^n is $\widehat{A}^*(\Delta^n) = A_*(\Delta^n)^\vee$, regarded as a commutative differential graded algebra.

Each of these constructions inherits a (co)simplicial structure; indeed, each is obtained by applying a functor pointwise to a (co)simplicial object.

DEFINITION A.2. Write $\iota : \Delta \to \mathrm{Fun}(\Delta^{\mathrm{op}}, \mathrm{Set})$ for the Yoneda embedding.

(1) The functor of *Sullivan cochains* is the left Kan extension A^* in the diagram

$$\begin{array}{ccc} \Delta & \xrightarrow{A^*(\Delta^\bullet)} & \mathrm{Alg}_{\mathrm{Com}}(\mathcal{C}\mathrm{h}_{\mathbb{Q}})^{\mathrm{op}} \\ {\scriptstyle \iota}\downarrow & \nearrow_{A^* = \iota_! A^*(\Delta^\bullet)} & \\ \mathrm{Fun}(\Delta^{\mathrm{op}}, \mathrm{Set}) & & \end{array}$$

(resp. completed Sullivan cochains, $\widehat{A}^*$).

(2) The functor of *Sullivan chains* is the left Kan extension A_* in the diagram

$$\begin{array}{ccc} \Delta & \xrightarrow{A_*(\Delta^\bullet)} & \mathrm{Coalg}_{\mathrm{Com}}(\mathcal{C}\mathrm{h}_{\mathbb{Q}}) \\ {\scriptstyle \iota}\downarrow & \nearrow_{A_* = \iota_! A_*(\Delta^\bullet)} & \\ \mathrm{Fun}(\Delta^{\mathrm{op}}, \mathrm{Set}). & & \end{array}$$

The complex of Sullivan (co)chains on the topological space X is the value of the corresponding functor on the singular set of X.

Note the canonical isomorphism $\widehat{A}^* \cong A_*^\vee$; indeed, dualization is colimit preserving when viewed as a functor from cocommutative differential graded coalgebras to the opposite category of commutative differential graded algebras.

LEMMA A.3. *The map $A^* \to \widehat{A}^*$ induced by the composite $V_\bullet \to (V_\bullet^\vee)^\vee \to \widehat{A}^*(\Delta^\bullet)$ is a quasi-isomorphism.*

PROOF. By [**12**, Prop. 10.5], it suffices to check that the induced map $A^*(\Delta^n) \to \widehat{A}^*(\Delta^n)$ is a quasi-isomorphism for each $n \geq 0$. Since the unique map $0 \to V_n$ is a quasi-isomorphism, and since symmetric powers preserve quasi-isomorphisms over $\mathbb{Q}$, it follows that the unit of $A^*(\Delta^n)$ is a quasi-isomorphism. Since the linear dual of a quasi-isomorphism is also a quasi-isomorphism, it follows that the unique map $V_n^\vee \to 0$ is a quasi-isomorphism; therefore, after invoking the same property of symmetric powers a second time, it follows that the counit of $A_*(\Delta^n)$ is a quasi-isomorphism. By the same property of the linear dual, it now follows that the unit of $\widehat{A}^*(\Delta^n)$ is a quasi-isomorphism. Since the map in question is compatible with units, being a map of algebras, the claim follows. □

THEOREM A.4. *Let K be a simplicial set and write N_* for the functor of rational normalized chains. There is a natural quasi-isomorphism $A_*(K) \simeq N_*(K)$ of differential graded coalgebras.*

PROOF. The proof closely follows that of [**12**, Thm. 10.9]. At the level of simplices, the respective counits induce the maps

$$A_*(\Delta^\bullet) \leftarrow A_*(\Delta^\bullet) \otimes N_*(\Delta^\bullet) \to N_*(\Delta^\bullet)$$

of cosimplicial differential graded coalgebras. Since both are objectwise acyclic, these maps are quasi-isomorphisms. Kan extending, one obtains the zig-zag

$$A_* = \iota_! A_*(\Delta^\bullet) \leftarrow \iota_! \left(A_*(\Delta^\bullet) \otimes N_*(\Delta^\bullet)\right) \to \iota_! N_*(\Delta^\bullet) \cong N_*,$$

where the rightmost isomorphism is justified by the observations that colimits of coalgebras are computed at the level of chain complexes; that N_* is colimit-preserving; and that the identity functor is the left Kan extension of the Yoneda embedding along itself. One could argue directly that these maps are quasi-isomorphisms by appealing to a cosimplicial version of [**12**, Prop. 10.5]. Alternatively, consider the following commutative diagram of functors from simplicial sets to commutative differential graded algebras:

$$\begin{array}{ccccc}
\widehat{A}^* & \longrightarrow & \iota_!\left((A_*(\Delta^\bullet)\otimes N_*(\Delta^\bullet))^\vee\right) & \longleftarrow & N^* \\
\Big\uparrow & & \Big\uparrow & & \Big\| \\
 & & \iota_!(\widehat{A}^*(\Delta^\bullet)\otimes N^*(\Delta^\bullet)) & & \Big\| \\
\Big| & & \Big\uparrow & & \Big\| \\
A^* & \longrightarrow & \iota_!\left(A^*(\Delta^\bullet)\otimes N^*(\Delta^\bullet)\right) & \longleftarrow & N^*,
\end{array}$$

the top row is the linear dual of the zig-zag of interest, the maps in the bottom row are shown to be quasi-isomorphisms in [**12**, Thm. 10.9], and the lefthand vertical arrow is a quasi-isomorphism by Lemma A.3; therefore, since formation of the linear dual reflects quasi-isomorphisms, it suffices to show that the vertical maps in the middle column are quasi-isomorphisms. By [**12**, Prop. 10.5], it suffices to show that the maps

$$A^*(\Delta^n)\otimes N^*(\Delta^n)\to \widehat{A}^*(\Delta^n)\otimes N^*(\Delta^n)\to (A_*(\Delta^n)\otimes N_*(\Delta^n))^\vee$$

are quasi-isomorphisms for each $n\geq 0$. The first is the quasi-isomorphism of Lemma A.3 tensored with the identity. The second is the canonical map from the tensor product of the linear duals to the linear dual of the tensor product, which is a quasi-isomorphism on complexes with finite dimensional homology. □

COROLLARY A.5. *Let X be a topological space. There is a natural quasi-isomorphism*

$$A_*(X)\simeq C_*(X;\mathbb{Q})$$

of differential graded coalgebras and of oplax monoidal functors.

PROOF. Restricting the quasi-isomorphism of Theorem A.4 along the singular set functor, we obtain the desired quasi-isomorphism of coalgebras. To complete the proof, we need only recall that the coalgebra structure on singular chains determines the oplax monoidal structure via

$$C_*(X\times Y)\to C_*(X\times Y)\otimes C_*(X\times Y)\xrightarrow{C_*(\pi_1)\otimes C_*(\pi_2)} C_*(X)\otimes C_*(Y).$$

□

REMARK A.6. One imagines that our approach in this appendix translates essentially unchanged to the setting of [**36**], providing a symmetric model for integral singular chains. We have not checked the details.

Acknowledgments

The author thanks Najib Idrissi and Roberto Pagaria for helpful conversations related to this work and the anonymous referee for her feedback. This paper was written for the proceedings of the conference "Compactifications, Configurations, and Cohomology," held at Northeastern University in October of 2021. The author extends his heartfelt gratitude to the organizers, Peter Crooks and Alex Suciu, for the opportunity to speak, to write, and to be reminded that face-to-face interaction in mathematics is a precious, even indispensable, commodity. This work was supported by NSF grant DMS-1906174.

References

[1] Marcelo Aguiar and Swapneel Mahajan, *Monoidal functors, species and Hopf algebras*, CRM Monograph Series, vol. 29, American Mathematical Society, Providence, RI, 2010. With forewords by Kenneth Brown and Stephen Chase and André Joyal, DOI 10.1090/crmm/029. MR2724388

[2] Byung Hee An, Gabriel C. Drummond-Cole, and Ben Knudsen, *Edge stabilization in the homology of graph braid groups*, Geom. Topol. **24** (2020), no. 1, 421–469, DOI 10.2140/gt.2020.24.421. MR4080487

[3] Christin Bibby and Nir Gadish, *A generating function approach to new representation stability phenomena in orbit configuration spaces*, Trans. Amer. Math. Soc. Ser. B **10** (2023), 241–287, DOI 10.1090/btran/130. MR4546785

[4] L. Brantner, J. Hahn, and B. Knudsen, *The Lubin–Tate theory of configuration spaces: I*, `arXiv:1908.11321`.

[5] Thomas Church and Jordan S. Ellenberg, *Homology of FI-modules*, Geom. Topol. **21** (2017), no. 4, 2373–2418, DOI 10.2140/gt.2017.21.2373. MR3654111

[6] Thomas Church, Jordan S. Ellenberg, and Benson Farb, *FI-modules and stability for representations of symmetric groups*, Duke Math. J. **164** (2015), no. 9, 1833–1910, DOI 10.1215/00127094-3120274. MR3357185

[7] Andrew A. Cooper, Vin de Silva, and Radmila Sazdanovic, *On configuration spaces and simplicial complexes*, New York J. Math. **25** (2019), 723–744. MR4012568

[8] Brian Day, *On closed categories of functors*, Reports of the Midwest Category Seminar, IV, Lecture Notes in Mathematics, Vol. 137, Springer, Berlin, 1970, pp. 1–38. MR0272852

[9] Gabriel C. Drummond-Cole and Ben Knudsen, *Betti numbers of configuration spaces of surfaces*, J. Lond. Math. Soc. (2) **96** (2017), no. 2, 367–393, DOI 10.1112/jlms.12066. MR3708955

[10] Michael Eastwood and Stephen Huggett, *Euler characteristics and chromatic polynomials*, European J. Combin. **28** (2007), no. 6, 1553–1560, DOI 10.1016/j.ejc.2006.09.005. MR2339484

[11] J. Ellenberg, T. Tran, and C. Westerland, *Fox–Neuwirth–Fuks cells, quantum shuffle algebras, and Malle's conjecture for function fields*, `arXiv:1701.04541`.

[12] Yves Félix, Stephen Halperin, and Jean-Claude Thomas, *Rational homotopy theory*, Graduate Texts in Mathematics, vol. 205, Springer-Verlag, New York, 2001, DOI 10.1007/978-1-4613-0105-9. MR1802847

[13] Yves Félix and Jean-Claude Thomas, *Rational Betti numbers of configuration spaces*, Topology Appl. **102** (2000), no. 2, 139–149, DOI 10.1016/S0166-8641(98)00148-5. MR1741482

[14] John Francis and Dennis Gaitsgory, *Chiral Koszul duality*, Selecta Math. (N.S.) **18** (2012), no. 1, 27–87, DOI 10.1007/s00029-011-0065-z. MR2891861

[15] N. Gadish and L. Hainaut, *Configuration spaces on a wedge of spheres and Hochschild–Pirashvili homology*, `arXiv:2202.12494`.

[16] E. Getzler, *Resolving mixed Hodge modules on configuration spaces*, Duke Math. J. **96** (1999), no. 1, 175–203, DOI 10.1215/S0012-7094-99-09605-9. MR1663927

[17] Saul Glasman, *Day convolution for ∞-categories*, Math. Res. Lett. **23** (2016), no. 5, 1369–1385, DOI 10.4310/MRL.2016.v23.n5.a6. MR3601070

[18] Daniel Grayson, *Higher algebraic K-theory. II (after Daniel Quillen)*, Algebraic K-theory (Proc. Conf., Northwestern Univ., Evanston, Ill., 1976), Lecture Notes in Math., Vol. 551, Springer, Berlin, 1976, pp. 217–240. MR0574096

[19] Richard Hepworth, *On the edge of the stable range*, Math. Ann. **377** (2020), no. 1-2, 123–181, DOI 10.1007/s00208-020-01955-0. MR4099630

[20] Sean Howe, *Motivic random variables and representation stability I: Configuration spaces*, Algebr. Geom. Topol. **20** (2020), no. 6, 3013–3045, DOI 10.2140/agt.2020.20.3013. MR4185934

[21] Quoc P. Ho, *Homological stability and densities of generalized configuration spaces*, Geom. Topol. **25** (2021), no. 2, 813–912, DOI 10.2140/gt.2021.25.813. MR4251437

[22] Najib Idrissi, *The Lambrechts-Stanley model of configuration spaces*, Invent. Math. **216** (2019), no. 1, 1–68, DOI 10.1007/s00222-018-0842-9. MR3935037

[23] Ben Knudsen, *Betti numbers and stability for configuration spaces via factorization homology*, Algebr. Geom. Topol. **17** (2017), no. 5, 3137–3187, DOI 10.2140/agt.2017.17.3137. MR3704255

[24] Ben Knudsen, *Higher enveloping algebras*, Geom. Topol. **22** (2018), no. 7, 4013–4066, DOI 10.2140/gt.2018.22.4013. MR3890770

[25] B. Knudsen, J. Miller, and P. Tosteson, *Extremal stability for configuration spaces*, Math. Ann., 2022. To appear.

[26] Jean-Louis Loday and Bruno Vallette, *Algebraic operads*, Grundlehren der mathematischen Wissenschaften [Fundamental Principles of Mathematical Sciences], vol. 346, Springer, Heidelberg, 2012, DOI 10.1007/978-3-642-30362-3. MR2954392

[27] M. A. Mandell, J. P. May, S. Schwede, and B. Shipley, *Model categories of diagram spectra*, Proc. London Math. Soc. (3) **82** (2001), no. 2, 441–512, DOI 10.1112/S0024611501012692. MR1806878

[28] J. P. May, *The geometry of iterated loop spaces*, Lecture Notes in Mathematics, Vol. 271, Springer-Verlag, Berlin-New York, 1972. MR0420610

[29] Dusa McDuff, *Configuration spaces of positive and negative particles*, Topology **14** (1975), 91–107, DOI 10.1016/0040-9383(75)90038-5. MR358766

[30] Jeremy Miller and Jennifer C. H. Wilson, *Higher-order representation stability and ordered configuration spaces of manifolds*, Geom. Topol. **23** (2019), no. 5, 2519–2591, DOI 10.2140/gt.2019.23.2519. MR4019898

[31] Rohit Nagpal, Steven V. Sam, and Andrew Snowden, *Noetherianity of some degree two twisted skew-commutative algebras*, Selecta Math. (N.S.) **25** (2019), no. 1, Paper No. 4, 26, DOI 10.1007/s00029-019-0461-3. MR3907946

[32] Dan Petersen, *Cohomology of generalized configuration spaces*, Compos. Math. **156** (2020), no. 2, 251–298, DOI 10.1112/s0010437x19007747. MR4045070

[33] Daniel Quillen, *Rational homotopy theory*, Ann. of Math. (2) **90** (1969), 205–295, DOI 10.2307/1970725. MR258031

[34] Eric Ramos, *Generalized representation stability and* FI_d*-modules*, Proc. Amer. Math. Soc. **145** (2017), no. 11, 4647–4660, DOI 10.1090/proc/13618. MR3691984

[35] Oscar Randal-Williams and Nathalie Wahl, *Homological stability for automorphism groups*, Adv. Math. **318** (2017), 534–626, DOI 10.1016/j.aim.2017.07.022. MR3689750

[36] Birgit Richter and Steffen Sagave, *A strictly commutative model for the cochain algebra of a space*, Compos. Math. **156** (2020), no. 8, 1718–1743, DOI 10.1112/s0010437x20007319. MR4163860

[37] Andrew Snowden, *Syzygies of Segre embeddings and* Δ*-modules*, Duke Math. J. **162** (2013), no. 2, 225–277, DOI 10.1215/00127094-1962767. MR3018955

[38] Burt Totaro, *Configuration spaces of algebraic varieties*, Topology **35** (1996), no. 4, 1057–1067, DOI 10.1016/0040-9383(95)00058-5. MR1404924

Department of Mathematics, Northeastern University, Boston, Massachusetts, 02115

Email address: b.knudsen@northeastern.edu

Contemporary Mathematics
Volume **790**, 2023
https://doi.org/10.1090/conm/790/15861

Cohomology, Bocksteins, and resonance varieties in characteristic 2

Alexandru I. Suciu

Abstract. We use the action of the Bockstein homomorphism on the cohomology ring $H^\bullet(X, \mathbb{Z}_2)$ of a finite-type CW-complex X in order to define the resonance varieties of X in characteristic 2. Much of the theory is done in the more general framework of the Maurer–Cartan sets and the resonance varieties attached to a finite-type commutative differential graded algebra. We illustrate these concepts with examples mainly drawn from closed manifolds, where Poincaré duality over $\mathbb{Z}_2$ has strong implications on the nature of the resonance varieties.

Contents

1. Introduction

1.1. Resonance varieties. The cohomology ring of a space X captures deep, albeit incomplete information about the homotopy type of X. A fruitful idea, which arose in the late 1990s from the theory of hyperplane arrangements [**8**], [**4**], [**16**], turns the cohomology algebra $H^\bullet(X, \Bbbk)$ over a coefficient field $\Bbbk$ into a family of cochain complexes with differentials given by multiplication by an element in the

---

2020 *Mathematics Subject Classification.* Primary 16E45, 55U20; Secondary 06E30, 14M12, 55S10, 55U30, 57P10.

Key words and phrases. Differential graded algebra, Maurer–Cartan set, Koszul complex, resonance variety, cohomology ring, Bockstein homomorphism, Poincaré duality, orientability, finite covers.

Supported in part by Simons Foundation Collaboration Grants for Mathematicians #354156 and #693825. ORCID 0000-0002-5060-7754.

vector space $H^1(X,\Bbbk)$. One extracts from these data the *resonance varieties* of X over $\Bbbk$, as the loci where the cohomology of the aforementioned cochain complexes jumps in a certain degree, by a specified amount.

Originally, the ground field was $\mathbb{C}$ and the space X was the complement of an arrangement of n hyperplanes in $\mathbb{C}^\ell$, in which case each of these resonance varieties was shown to be a finite union of linear subspaces in $\mathbb{C}^n$. Starting with [**20**], the scope of investigation of the resonance varieties broadened, with fields $\Bbbk$ of arbitrary characteristic and more general spaces X being allowed, as long as X had the homotopy type of a connected, finite-type CW-complex, its homology groups $H_\bullet(X,\mathbb{Z})$ were free abelian, and the cohomology ring $H^\bullet(X,\mathbb{Z})$ was generated in degree 1. New phenomena were discovered in this wider generality; for instance, the resonance varieties over $\mathbb{C}$ no longer need to be linear, while the resonance varieties over $\mathbb{Z}_p$ (for p a prime) carry useful information regarding the index-p subgroups of the second nilpotent quotient of $\pi_1(X)$ (see also [**31**]). Further investigations of the resonance varieties over finite fields were done in [**9**].

The theory was developed in even greater generality in [**23**] and [**24**], where the only condition required was that either $\operatorname{char}(\Bbbk) \neq 2$, or $\operatorname{char}(\Bbbk) = 2$ and $H_1(X,\mathbb{Z})$ to be torsion-free, so as to insure that $a^2 = 0$ for every $a \in H^1(X,\Bbbk)$. Unfortunately, this torsion-freeness condition rules out many interesting spaces. In this note, we remedy this situation, by redefining the resonance varieties in characteristic 2, so as to take into account the action of the Bockstein operator, $\beta_2 = \operatorname{Sq}^1$, on the cohomology algebra $H^\bullet(X,\mathbb{Z}_2)$.

1.2. Commutative differential graded algebras. The key idea is to appeal to a construction done in an even wider framework. This construction, which was first introduced and studied in [**18**], [**25**], and [**34**], and is being further developed in [**5**], [**36**], and [**37**], starts with a commutative differential graded algebra (A,d). For every element a in the *Maurer–Cartan set* of this CDGA,

$$\mathrm{MC}(A) = \{a \in A^1 \mid a^2 + \mathrm{d}(a) = 0 \in A^2\}, \tag{1}$$

one constructs a cochain complex $(A^\bullet, \delta_a)$ with differentials $\delta^i_a\colon A^i \to A^{i+1}$ the $\Bbbk$-linear maps given by $\delta^i_a(u) = a\cdot u + \mathrm{d}(u)$ for $u \in A^i$. The resonance varieties of A (in degree $q \geq 0$ and depth $s \geq 0$) are the loci

$$\mathcal{R}^q_s(A) = \big\{a \in \mathrm{MC}(A) \mid \dim_\Bbbk H^q(A,\delta_a) \geq s\big\}. \tag{2}$$

Under the assumption that all the graded pieces A^i are finite-dimensional, these sets are Zariski closed subsets of the algebraic variety $\mathrm{MC}(A)$; when A has zero differential, the resonance varieties are homogeneous, but in general that is not the case.

Especially interesting is the case when (A,d) is a Poincaré duality CDGA of formal dimension m, that is, the underlying graded algebra $A^\bullet$ satisfies Poincaré duality and the differential d vanishes on A^{m-1}. Building on work from [**27**] and [**35**], we show in Theorem 4.6 that, for such CDGAs, the involution $a \mapsto -a$ on A^1 restricts to isomorphisms

$$\mathcal{R}^q_s(A) \xrightarrow{\ \simeq\ } \mathcal{R}^{m-q}_s(A) \tag{3}$$

for all $q, s \geq 0$.

1.3. The Bockstein operator and resonance in characteristic 2. We apply the general theory outlined above to the case when $A = H^{\bullet}(X, \mathbb{Z}_2)$ is the cohomology algebra of a connected, finite-type CW-complex X, equipped with the differential $\beta_2 \colon A^{\bullet} \to A^{\bullet+1}$ given by the Bockstein homomorphism associated to the coefficient exact sequence $0 \to \mathbb{Z}_2 \to \mathbb{Z}_4 \to \mathbb{Z}_2 \to 0$. We denote the resonance varieties of the CDGA (A, β_2) by $\widetilde{\mathcal{R}}^q_s(X, \mathbb{Z}_2)$. Slightly more generally, for a field $\Bbbk$ of characteristic 2, we define the *Bockstein resonance varieties* of X as

$$\widetilde{\mathcal{R}}^q_s(X, \Bbbk) = \widetilde{\mathcal{R}}^q_s(X, \mathbb{Z}_2) \times_{\mathbb{Z}_2} \Bbbk. \tag{4}$$

If the group $H_1(X, \mathbb{Z})$ has no 2-torsion, one may also define the "usual" resonance varieties, $\mathcal{R}^q_s(X, \Bbbk)$, as the resonance varieties of the CDGA $(A, 0)$. It turns out that $\mathcal{R}^1_s(X, \Bbbk) = \widetilde{\mathcal{R}}^1_s(X, \Bbbk)$ for all $s \geq 0$, due to the vanishing of the Bockstein $\beta_2 \colon H^1(X, \mathbb{Z}_2) \to H^2(X, \mathbb{Z}_2)$ in this situation. Nevertheless, the resonance varieties $\mathcal{R}^q_s(X, \Bbbk)$ and $\widetilde{\mathcal{R}}^q_s(X, \Bbbk)$ may differ in degrees $q > 1$; in fact, we give examples showing that either variety may be properly included in the other when $s = 1$. For instance, if M is a smooth, closed, non-orientable manifold of dimension m and $H_1(M, \mathbb{Z})$ has no 2-torsion, we then show in Corollary 7.11 that $\mathcal{R}^m_1(M, \mathbb{Z}_2) = \{0\}$, whereas $\widetilde{\mathcal{R}}^m_1(M, \mathbb{Z}_2) = \mathbb{Z}_2$.

1.4. Poincaré duality, orientability, and resonance. In fact, much more can be said in the case when our space is a topological manifold M. We always assume M to be a closed manifold (that is, compact, connected, with no boundary). Our first main result in this context (proved in Corollary 7.3 and Proposition 7.10) characterizes the orientability of M in terms of its Bockstein CDGA and its top-degree Bockstein resonance variety.

THEOREM 1.1. *Let M be a closed, smooth manifold of dimension m. The following conditions are equivalent.*

(1) *M is orientable.*
(2) *$(H^{\bullet}(M, \mathbb{Z}_2), \beta_2)$ is a Poincaré duality differential graded algebra (of formal dimension m).*
(3) $\widetilde{\mathcal{R}}^m_1(M, \mathbb{Z}_2) = \{0\}$.

Our second main result (proved in Propositions 7.5 and 7.7) is a Poincaré duality analog for the resonance varieties (of both types) of closed, orientable manifolds.

THEOREM 1.2. *Let M be a closed, orientable manifold of dimension m.*

(1) *If $\operatorname{char}(\Bbbk) \neq 2$, then $\mathcal{R}^q_s(M; \Bbbk) = \mathcal{R}^{m-q}_s(M; \Bbbk)$ for all $q, s \geq 0$.*
(2) *If $\operatorname{char}(\Bbbk) = 2$, then $\widetilde{\mathcal{R}}^q_s(M; \Bbbk) = \widetilde{\mathcal{R}}^{m-q}_s(M; \Bbbk)$ for all $q, s \geq 0$.*

We illustrate these results with several classes of manifolds, including the orientable and non-orientable surfaces, some lens spaces and other 3-manifolds, the real projective spaces $\mathbb{RP}^n$, and the Dold manifolds $P(m, n)$.

1.5. Betti numbers in finite covers. In the final section we discuss some of the ways in which the resonance varieties $\mathcal{R}^q_s(X, \Bbbk)$ and $\widetilde{\mathcal{R}}^q_s(X, \Bbbk)$, as well as the Betti numbers $b_q(X, \Bbbk) = \dim_{\Bbbk} H^q(X, \Bbbk)$ behave when passing to finite covers.

For instance, suppose $Y \to X$ is a connected 2-fold cover classified by a nonzero class $\alpha \in H^1(X, \mathbb{Z}_2)$ with $\alpha^2 = 0$. In Proposition 8.5 we strengthen results from [**39**] and show that

$$b_q(Y,\mathbb{Z}_2) = b_q(X,\mathbb{Z}_2) + \dim_{\mathbb{Z}_2} H^q(H^\bullet(X,\mathbb{Z}_2),\delta_\alpha), \tag{5}$$

for all $q \geq 1$. In particular, the mod-2 Betti numbers of Y are at least as large as those of X, which is not necessarily the case when $\alpha^2 \neq 0$.

2. Maurer–Cartan elements and Koszul complexes

We start with the Maurer–Cartan sets and the Koszul complexes associated to a commutative differential graded algebra.

2.1. Commutative graded algebras. Throughout this work, $\Bbbk$ will denote a ground field. A *graded $\Bbbk$-vector space* is a vector space A over $\Bbbk$, together with a direct sum decomposition into vector subspaces, $A = \bigoplus_{i\geq 0} A^i$, called its graded pieces. An element $a \in A^i$ is said to be homogeneous; we write $|a| = i$ for its degree.

A *graded algebra* over $\Bbbk$ is a graded $\Bbbk$-vector space, $A^\bullet = \bigoplus_{i\geq 0} A^i$, equipped with an associative multiplication map, $\cdot\colon A \times A \to A$, making A into a $\Bbbk$-algebra with unit $1 \in A^0$ such that $|a\cdot b| = |a| + |b|$ for all homogenous elements $a, b \in A$. A graded algebra A is said to be *graded-commutative* (for short, a CGA), if $a \cdot b = (-1)^{|a||b|} b \cdot a$ for all homogeneous $a, b \in A$, and *strictly commutative* if $ab = ba$ for all $a, b \in A$. Note that the two versions of commutativity agree when $\operatorname{char}(\Bbbk) = 2$.

A morphism between two graded algebras is a $\Bbbk$-linear map $\varphi\colon A \to B$ such that φ sends A^i to B^i for all $i \geq 0$ and satisfies $\varphi(a\cdot b) = \varphi(a)\cdot\varphi(b)$ for all $a, b \in A$.

We say that a graded algebra A is *connected* if A^0 is the $\Bbbk$-span of the unit 1 (and thus $A^0 = \Bbbk$). We also say that A is of *finite-type* (or, locally finite) if all the graded pieces A^i are finite-dimensional.

Given a vector space V, one may construct several basic examples of graded algebras from it: the tensor algebra $T = T(V)$, the symmetric algebra $S = \operatorname{Sym}(V)$, and the exterior algebra $E = \bigwedge(V)$. All three algebras are connected; moreover, E is graded-commutative, S is strictly commutative, and T has neither of these properties. If V is finite-dimensional, then T and S are of finite-type and E is finite-dimensional.

2.2. Commutative differential graded algebras. We now enrich the notion of a CGA by introducing a differential, which we require to be compatible with both the grading and the multiplication. By definition, a *commutative differential graded algebra* over $\Bbbk$ (for short, a $\Bbbk$-CDGA) is a pair $A = (A^\bullet, \mathrm{d})$, where $A^\bullet$ is a $\Bbbk$-CGA and $\mathrm{d}\colon A \to A$ is a *graded derivation*; that is, d is a $\Bbbk$-linear map such that $\mathrm{d}\circ\mathrm{d} = 0$, $\mathrm{d}(A^i) \subseteq A^{i+1}$ for all $i \geq 0$, and $\mathrm{d}(a\cdot b) = \mathrm{d}(a)\cdot b + (-1)^{|a|} a \cdot \mathrm{d}(b)$, for all homogeneous elements $a, b \in A$.

Using only the underlying cochain complex structure of the CDGA, we let $Z^i(A) = \ker(\mathrm{d}\colon A^i \to A^{i+1})$ be the subspace of cocycles and $B^i(A) = \operatorname{im}(\mathrm{d}\colon A^{i-1} \to A^i)$ the subspace of coboundaries, and define the i-th cohomology group of $(A^\bullet, \mathrm{d})$ as the quotient $\Bbbk$-vector space $H^i(A) = Z^i(A)/B^i(A)$. The direct sum of these vector spaces, $H^\bullet(A) = \bigoplus_{i\geq 0} H^i(A)$, inherits from A the structure of a graded, graded-commutative $\Bbbk$-algebra. If A is of finite-type, then clearly $H^\bullet(A)$ is also of finite-type; in this case, we define the Betti numbers of A to be the ranks of the graded pieces of $H^\bullet(A)$, and write them as $b_i(A) := \dim_\Bbbk H^i(A)$.

For a cocycle $z \in Z^i(A)$, we will denote by $[z] \in H^i(A)$ the cohomology class it represents. Observe that $\mathrm{d}(1) = \mathrm{d}(1\cdot 1) = \mathrm{d}(1) + \mathrm{d}(1)$, and so $\mathrm{d}(1) = 0$. Therefore, if A is connected, the differential $\mathrm{d}\colon A^0 \to A^1$ vanishes, and so we may identify $H^0(A) = \Bbbk$ and $H^1(A) = Z^1(A)$.

A morphism between two CDGAs, $\varphi\colon (A, \mathrm{d}_A) \to (B, \mathrm{d}_B)$, is both a map of graded algebras and a cochain map; that is, φ is a $\Bbbk$-linear map that preserves gradings, multiplicative structures, and differentials. Denoting by $\varphi^i\colon A^i \to B^i$ the restriction of φ to i-th graded pieces, we have that $\varphi^i(ab) = \varphi^i(a)\varphi^i(b)$ and $\mathrm{d}_B(\varphi^i(a)) = \varphi^{i+1}(\mathrm{d}_A(a))$, for all $a, b \in A^i$. The morphism φ induces a morphism $\varphi^*\colon H^\bullet(A) \to H^\bullet(B)$ between the respective cohomology algebras. We say that φ is a *quasi-isomorphism* if φ^* is an isomorphism.

A *weak equivalence* between two CDGAs, A and B, is a finite sequence of quasi-isomorphisms going either way and connecting A to B; one such zig-zag of quasi-isomorphisms is given in the diagram below,

$$A \xleftarrow{\ \varphi_1\ } A_1 \xrightarrow{\ \varphi_2\ } \cdots \longleftarrow A_{\ell-1} \xrightarrow{\ \varphi_\ell\ } B. \tag{6}$$

Note that a weak equivalence induces a well-defined isomorphism $H^\bullet(A) \cong H^\bullet(B)$. If a weak equivalence between A and B exists, we say that the two CDGAs are *weakly equivalent*, and write $A \simeq B$. Clearly, $\simeq$ is an equivalence relation among CDGAs.

2.3. Maurer–Cartan sets. Let (A, d) be a CDGA over a field $\Bbbk$. We define the *Maurer–Cartan set* of A by

$$\mathrm{MC}(A) = \left\{a \in A^1 \mid a^2 + \mathrm{d}(a) = 0 \in A^2\right\}. \tag{7}$$

If A^1 is finite-dimensional, then the set of Maurer–Cartan elements is a quadratic algebraic subvariety of the affine space A^1. Clearly, this variety contains the point $0 \in A^1$. Two general classes of examples are worth singling out.

EXAMPLE 2.1. Suppose $a^2 + \mathrm{d}(a) = 0$ for all $a \in A^1$. In this case (which we will concentrate on in the latter sections), we clearly have $\mathrm{MC}(A) = A^1$.

EXAMPLE 2.2. Suppose $a^2 = 0$, for every $a \in A^1$—a condition that is always satisfied if $\mathrm{char}(\Bbbk) \neq 2$. In this case, $\mathrm{MC}(A) = Z^1(A)$. If, additionally, A is connected, then a previous observation gives $\mathrm{MC}(A) = H^1(A)$.

In all the above examples, the Maurer–Cartan set is a linear subspace of A^1. In general, though, $\mathrm{MC}(A)$ is cut out by quadratic and linear equations; here is a simple example illustrating this fact.

EXAMPLE 2.3. Let $A = (\mathbb{Z}_2[a_1, a_2], \mathrm{d})$, with differential given by $\mathrm{d}(a_1) = a_1a_2$ and $\mathrm{d}(a_2) = a_2^2$. Then $\mathrm{MC}(A) = \{(x_1, x_2) \in \mathbb{Z}_2^2 \mid x_1(x_1 + x_2) = 0\}$.

The next lemma (which shall be of use in §4.2) shows that the Maurer–Cartan set in invariant under negation in the ambient vector space.

LEMMA 2.4. *Let (A, d) be a $\Bbbk$-CDGA. Then the linear involution $A^1 \to A^1$, $a \mapsto -a$ restricts to an involution $\mathrm{MC}(A) \to \mathrm{MC}(A)$.*

PROOF. If $\mathrm{char}(\Bbbk) = 2$, then the claim is obviously true. On the other hand, if $\mathrm{char}(\Bbbk) \neq 2$, then $\mathrm{MC}(A) = Z^1(A)$, which is a linear subspace of A^1, and we are done. □

The construction of the Maurer–Cartan sets is functorial. More precisely, we have the following lemma, which follows directly from the definitions.

LEMMA 2.5. *Let* $\varphi\colon A \to B$ *be a morphism of* CDGAs. *Then the linear map* $\varphi^1\colon A^1 \to B^1$ *restricts to a map,* $\bar{\varphi}\colon \mathrm{MC}(A) \to \mathrm{MC}(B)$, *between the respective subsets; moreover,* $\overline{\psi \circ \varphi} = \bar{\psi} \circ \bar{\varphi}$.

If both A^1 and B^1 are finite-dimensional, then the map $\bar{\varphi}\colon \mathrm{MC}(A) \to \mathrm{MC}(B)$ from above is a morphism of algebraic varieties. In particular, if $\varphi\colon A \to B$ is an isomorphism, then $\bar{\varphi}$ is also an isomorphism.

It also follows from the definitions that the construction is compatible with restriction and extension of the base field.

LEMMA 2.6. *Let* A *be a* $\Bbbk$-CDGA *with* $\dim_{\Bbbk} A^1 < \infty$, *and let* $\Bbbk \subset \mathbb{K}$ *be a field extension. Then,*

(1) $\mathrm{MC}(A) = \mathrm{MC}(A \otimes_{\Bbbk} \mathbb{K}) \cap A^1$.
(2) $\mathrm{MC}(A \otimes_{\Bbbk} \mathbb{K}) = \mathrm{MC}(A) \times_{\Bbbk} \mathbb{K}$.

2.4. The cochain complex associated to a Maurer–Cartan element. Let (A, d) be a finite-type $\Bbbk$-CDGA. For each element $a \in \mathrm{MC}(A)$, we have a cochain complex of finite-dimensional $\Bbbk$-vector spaces,

$$(A^{\bullet}, \delta^A_a)\colon\ A^0 \xrightarrow{\delta^0_a} A^1 \xrightarrow{\delta^1_a} A^2 \xrightarrow{\delta^2_a} \cdots, \tag{8}$$

with differentials $\delta^i_a\colon A^i \to A^{i+1}$ the $\Bbbk$-linear maps given by

$$\delta^i_a(u) = a \cdot u + \mathrm{d}(u), \tag{9}$$

for $u \in A^i$. The fact that these maps are differentials is readily verified:

$$\begin{aligned} \delta^{i+1}_a \delta^i_a(u) &= a^2 u + a \cdot \mathrm{d}(u) + \mathrm{d}(a) \cdot u - a \cdot \mathrm{d}(u) + \mathrm{d}(\mathrm{d}(u)) \\ &= (a^2 + \mathrm{d}(a)) \cdot u \\ &= 0, \end{aligned} \tag{10}$$

where at the last step we used the assumption that a belongs to $\mathrm{MC}(A)$. We let

$$b_i(A, a) := \dim_{\Bbbk} H^i(A, \delta^A_a) \tag{11}$$

be the Betti numbers of this cochain complex. Observe that $\delta_0 = \mathrm{d}$, and thus $b_i(A, 0) = b_i(A)$.

The cochain complex associated to a Maurer–Cartan element enjoys the following naturality property. Related statements (in different levels of generality) can be found in [**5**] and [**35**].

LEMMA 2.7. *Let* $\varphi\colon (A, \mathrm{d}_A) \to (B, \mathrm{d}_B)$ *be a morphism of finite-type* $\Bbbk$-CDGAs. *For each* $a \in \mathrm{MC}(A)$, *the map* φ *induces a chain map,*

$$\begin{array}{ccccccccccc} (A^{\bullet}, \delta^A_a): & A^0 & \xrightarrow{\delta^0_a} & A^1 & \longrightarrow \cdots \longrightarrow & A^i & \xrightarrow{\delta^i_a} & A^{i+1} & \xrightarrow{\delta^{i+1}_a} & \cdots \\ \downarrow{\scriptstyle \varphi_a} & \downarrow{\scriptstyle \varphi^0} & & \downarrow{\scriptstyle \varphi^1} & & \downarrow{\scriptstyle \varphi^i} & & \downarrow{\scriptstyle \varphi^{i+1}} & & \\ \left(B^{\bullet}, \delta^B_{\varphi(a)}\right): & B^0 & \xrightarrow{\delta^0_{\varphi(a)}} & B^1 & \longrightarrow \cdots \longrightarrow & B^i & \xrightarrow{\delta^i_{\varphi(a)}} & B^{i+1} & \xrightarrow{\delta^{i+1}_{\varphi(a)}} & \cdots. \end{array} \tag{12}$$

PROOF. Let $u \in A^i$. By definition, $\varphi_a(u) = \varphi^i(u)$. Hence,

$$\begin{aligned} \varphi_a(\delta^A_a(u)) &= \varphi^{i+1}(au + \mathrm{d}_A(u)) \\ &= \varphi^1(a)\varphi^i(u) + \mathrm{d}_B(\varphi^i(u)) \\ &= \delta^B_{\varphi(a)}(\varphi_a(u)), \end{aligned} \tag{13}$$

and the claim is proved. □

Consequently, for each $a \in \mathrm{MC}(A)$, the chain map $\varphi_a \colon (A^{\bullet}, \delta^A_a) \to (B^{\bullet}, \delta^B_{\varphi(a)})$ induces homomorphisms in cohomology,

$$\varphi^i_a \colon H^i\left(A, \delta^A_a\right) \longrightarrow H^i\left(B, \delta^B_{\bar{\varphi}(a)}\right). \tag{14}$$

LEMMA 2.8. *Let $\varphi \colon (A, \mathrm{d}_A) \to (B, \mathrm{d}_B)$ be a morphism of finite-type $\Bbbk$-CDGAs, let $a \in \mathrm{MC}(A)$, and let i be a positive integer.*

(1) *Suppose φ^i is injective and φ^{i-1} is surjective. Then φ^i_a is injective.*
(2) *Suppose φ^i is surjective and φ^{i+1} is injective. Then φ^i_a is surjective.*

PROOF. (1) Suppose $\varphi^i_a([u]) = 0$, for some $u \in A^i$ with $au + \mathrm{d}_A(u) = 0$. Then $\varphi^i(u) = \varphi^1(a)v + \mathrm{d}_B(v)$, for some $v \in B^{i-1}$. By our surjectivity assumption on φ^{i-1}, there is an element $w \in A^{i-1}$ such that $\varphi^{i-1}(w) = v$, and so $\varphi^i(u) = \varphi^i(aw) + \varphi^i(\mathrm{d}_A(w))$. Our injectivity assumption on φ^i now implies that $u = aw + \mathrm{d}_A(w)$, and so $[u] = 0$.

(2) Let $[v] \in H^i(B, \delta^B_{\bar{\varphi}(a)})$, for some $v \in B^i$ with $\varphi^1(a)v + \mathrm{d}_B(v) = 0$. Since φ^i is surjective, there is an element $u \in A^i$ such that $v = \varphi^i(u)$. We then have

$$\begin{aligned} \varphi^{i+1}(au + \mathrm{d}_A(u)) &= \varphi^1(a)\varphi^i(u) + \mathrm{d}_B(\varphi^i(u)) \\ &= \varphi^1(a)v + \mathrm{d}_B(v) \\ &= 0, \end{aligned} \tag{15}$$

and, since φ^{i+1} is injective, we have that $au + \mathrm{d}_A(u) = 0$. Therefore, $[v] = \varphi^i_a([u])$, and the proof is complete. □

2.5. A generalized Koszul complex. We now fix a basis $\{e_1, \dots, e_n\}$ for the finite-dimensional $\Bbbk$-vector space A^1, and we let $\{x_1, \dots, x_n\}$ be the Kronecker dual basis for the dual vector space $A_1 = (A^1)^{\vee}$. In what follows, we shall identify the symmetric algebra $\mathrm{Sym}(A_1)$ with the polynomial ring $R = \Bbbk[x_1, \dots, x_n]$.

The coordinate ring of the affine variety $\mathrm{MC}(A) \subset A^1$ is the quotient, $S = R/I$, of the ring R by the defining ideal of $\mathrm{MC}(A)$. Consider the cochain complex of free S-modules,

(16)

$$(A^{\bullet} \otimes_{\Bbbk} S, \delta_A) \colon \cdots \longrightarrow A^i \otimes_{\Bbbk} S \xrightarrow{\delta^i} A^{i+1} \otimes_{\Bbbk} S \xrightarrow{\delta^{i+1}} A^{i+2} \otimes_{\Bbbk} S \longrightarrow \cdots,$$

where the differentials $\delta^i = \delta^i_A$ are the S-linear maps defined by

$$\delta^i(u \otimes s) = \sum_{j=1}^{n} e_j u \otimes s x_j + \mathrm{d}(u) \otimes s \tag{17}$$

for all $u \in A^i$ and $s \in S$. The fact that this is a cochain complex is easily verified; indeed, $\delta^{i+1}\delta^i(u \otimes s)$ is equal to

$$
\begin{aligned}
&\sum_k e_k\Big(\sum_j e_j u \otimes s x_j + \mathrm{d}\, u \otimes s\Big) \otimes x_k + \mathrm{d}\Big(\sum_j e_j u \otimes s x_j + \mathrm{d}\, u \otimes s\Big) \\
&\qquad = \sum_{j,k} e_k e_j u \otimes s x_j x_k + \sum_k e_k\, \mathrm{d}\, u \otimes s x_k - \sum_j e_j\, \mathrm{d}\, u \otimes s x_j \\
&\qquad = 0,
\end{aligned} \tag{18}
$$

where at the last step we used the fact that $e_k e_j = -e_j e_k$.

REMARK 2.9. The cochain complex (16) is independent of the choice of basis for the vector space A^1. Indeed, under the canonical identification $A^1 \otimes_{\Bbbk} A_1 \cong \operatorname{Hom}(A^1, A^1)$, the element $\iota = \sum_{j=1}^n e_j \otimes x_j$ used in defining the differentials δ^i corresponds to the identity map of A^1.

EXAMPLE 2.10. Let $E = \bigwedge(e_1, \dots, e_n)$ be an exterior algebra over a field $\Bbbk$ of characteristic not equal to 2 (with generators e_i in degree 1 and with zero differential), and let $S = \Bbbk[x_1, \dots, x_n]$ be its Koszul dual. Then the cochain complex $(E^{\bullet} \otimes_{\Bbbk} S, \delta)$ is the classical Koszul complex $K_{\bullet}(x_1, \dots, x_n)$.

More generally, if the CDGA A has zero differential, each boundary map $\delta^i \colon A^i \otimes_{\Bbbk} S \to A^{i+1} \otimes_{\Bbbk} S$ is given by a matrix whose entries are linear forms in the variables $x_1, \dots, x_n$. If the differential of A is non-zero, though, the entries of δ^i may also have non-zero constant terms. We give a simple example, extracted from [**18**] and [**34**].

EXAMPLE 2.11. Let $\bigwedge(a, b)$ be the exterior algebra on generators a, b in degree 1 over a field $\Bbbk$ with $\operatorname{char}(\Bbbk) \neq 2$, and let $A = (\bigwedge(a, b), \mathrm{d})$ be the CDGA with differential given by $\mathrm{d}(a) = 0$ and $\mathrm{d}(b) = b \cdot a$. Then $\operatorname{MC}(A) = H^1(A)$ is 1-dimensional, generated by a. Writing $S = \Bbbk[x]$, the cochain complex (16) takes the form

$$
S \xrightarrow{\ \delta^0 = (\, x \ 0\,)\ } S^2 \xrightarrow{\ \delta^1 = \left(\begin{smallmatrix} 0 \\ x-1 \end{smallmatrix}\right)\ } S. \tag{19}
$$

The relationship between the cochain complexes (8) and (16) is given by the following lemma, which is a slight generalization of a known result (see e.g. [**34**]).

LEMMA 2.12. *The specialization of the cochain complex* $(A \otimes_{\Bbbk} S, \delta)$ *at an element* $a \in \operatorname{MC}(A)$ *coincides with the cochain complex* (A, δ_a).

PROOF. Write $a = \sum_{j=1}^n a_j e_j \in A^1$, and let $\mathfrak{m}_a = (x_1 - a_1, \dots, x_n - a_n)$ be the maximal ideal at a. The evaluation map $\operatorname{ev}_a \colon S \to S/\mathfrak{m}_a = \Bbbk$ is the ring morphism given by $g \mapsto g(a_1, \dots, a_n)$. The resulting cochain complex, $A(a) = A \otimes_S S/\mathfrak{m}_a$, has differentials $\delta^i(a) \colon A^i \to A^{i+1}$ given by

$$
\begin{aligned}
\delta^i(a)(u) &= \sum_{j=1}^n e_j u \otimes \operatorname{ev}_a(x_j) + \mathrm{d}(u) \\
&= \sum_{j=1}^n e_j u \cdot a_j + \mathrm{d}(u) \\
&= a \cdot u + \mathrm{d}(u),
\end{aligned} \tag{20}
$$

for all $u \in A^i$. Thus, $A(a) = (A, \delta_a)$, as claimed. ∎

3. Resonance varieties of commutative differential graded algebras

In this section we study the resonance varieties associated to a finite-type commutative differential graded algebra over an arbitrary field.

3.1. Resonance varieties. Let (A, d) be a finite-type CDGA over a field $\Bbbk$. Computing the homology of the cochain complexes (A, δ_a) for various values of the parameter $a \in \mathrm{MC}(A)$ and keeping track of the resulting Betti numbers carves out noteworthy subsets of the Maurer–Cartan set of A.

DEFINITION 3.1. The *resonance varieties* (in degree $q \geq 0$ and depth $s \geq 0$) of a finite-type $\Bbbk$-CDGA (A, d) are the sets

$$\mathcal{R}^q_s(A) = \{a \in \mathrm{MC}(A) \mid \dim_{\Bbbk} H^q(A, \delta_a) \geq s\}. \tag{21}$$

As we shall see below, these sets are, in fact, Zariski closed subsets of $\mathrm{MC}(A)$. Note that, for a fixed $q \geq 0$, the degree-q resonance varieties form a descending filtration of the Maurer–Cartan set,

$$\mathrm{MC}(A) = \mathcal{R}^q_0(A) \supseteq \mathcal{R}^q_1(A) \supseteq \mathcal{R}^q_2(A) \supseteq \cdots . \tag{22}$$

Here is a more concrete description of these varieties, which follows at once from the definitions.

LEMMA 3.2. *Fix integers $q \geq 1$ and $s \geq 0$. An element $a \in \mathrm{MC}(A)$ belongs to $\mathcal{R}^q_s(A)$ if and only if there exist elements $u_1, \dots, u_s \in A^q$ such that $au_1 + \mathrm{d}(u_1) = \cdots = au_s + \mathrm{d}(u_s) = 0$ in A^{q+1}, and the set $\{av + \mathrm{d}(v), u_1, \dots, u_s\}$ is linearly independent in A^q, for all $v \in A^{q-1}$.*

When the algebra A is connected, the differential $\mathrm{d}\colon A^0 \to A^1$ vanishes, and the next lemma readily follows.

LEMMA 3.3. *Let (A, d) be a connected, finite-type $\Bbbk$-CDGA. Then,*

1. *The point $0 \in \mathrm{MC}(A)$ belongs to $\mathcal{R}^q_s(A)$ if and only if $s \leq b_q(A)$.*
2. *$\mathcal{R}^0_1(A) = \{0\}$ and $\mathcal{R}^0_s(A) = \emptyset$ for $s > 1$.*
3. *A non-zero element $a \in \mathrm{MC}(A)$ belongs to $\mathcal{R}^1_1(A)$ if and only if there is an element $u \in A^1$ not proportional to a and satisfying $au + \mathrm{d}(u) = 0$.*
4. *If $\dim_{\Bbbk} A^1 = 1$, then $\mathcal{R}^1_1(A) = \{0\}$ if $b_1(A) = 1$ and $\mathcal{R}^1_1(A) = \emptyset$ if $b_1(A) = 0$.*

3.2. Equations for the resonance varieties. Once again, let (A, d) be a finite-type $\Bbbk$-CDGA. By definition, an element $a \in \mathrm{MC}(A)$ belongs to $\mathcal{R}^q_s(A)$ if and only if

$$\operatorname{rank} \delta^{q-1}_a + \operatorname{rank} \delta^q_a \leq c_q - s, \tag{23}$$

where $c_q = \dim_{\Bbbk} A^q$. Let $R = \Bbbk[x_1, \dots, x_n]$, let $S = R/I$ be the coordinate ring of $\mathrm{MC}(A)$, and let $(A^{\bullet} \otimes_{\Bbbk} S, \delta_A)$ be the cochain complex of free, finitely generated S-modules from (16). It follows from Lemma 2.12 that

$$\mathcal{R}^q_s(A) = V\Big(I_{c_q - s + 1}\big(\delta^{q-1}_A \oplus \delta^q_A\big)\Big). \tag{24}$$

Here, $\oplus$ denotes block-sum of matrices; for an $m \times n$ matrix ψ with entries in S, we let $I_r(\psi)$ denote the ideal of $r \times r$ minors of ψ, with the convention that $I_0(\psi) = S$ and $I_r(\psi) = 0$ if $r > \min(m, n)$; finally, $V(\mathfrak{a}) \subset \mathrm{MC}(A)$ denotes the zero set of an ideal $\mathfrak{a} \subset S$. This shows that the resonance sets $\mathcal{R}^q_s(A)$ are indeed subvarieties of the Maurer–Cartan variety $\mathrm{MC}(A) \subset A^1$.

EXAMPLE 3.4. Let (A, d) be the CDGA with $A = \mathbb{Z}_2[a_1, a_2]/(a_1^2, a_2^3, a_1a_2)$, where $|a_i| = 1$ and with differential given by $\mathrm{d}(a_1) = 0$, $\mathrm{d}(a_2) = a_2^2$. Then $\mathrm{MC}(A)$ is equal to $A^1 = \mathbb{Z}_2^2$, and it follows that $S = \mathbb{Z}_2[x_1, x_2]/(x_1^2 + x_1, x_2^2 + x_2)$, see the discussion from §5.2 below. The chain complex $(A \otimes_{\mathbb{Z}_2} S, \delta)$ now takes the form

$$S \xrightarrow{\delta^0 = (\,x_1 \;\; x_2\,)} S^2 \xrightarrow{\delta^1 = \left(\begin{smallmatrix} 0 \\ x_2+1 \end{smallmatrix}\right)} S,$$

By formula (24), the resonance varieties $\mathcal{R}^1_s(A)$ are the zero loci of the ideals of minors of size $3 - s$ of the matrix

$$\delta^0 \oplus \delta^1 = \begin{pmatrix} x_1 & x_2 & 0 \\ 0 & 0 & 0 \\ 0 & 0 & x_2 + 1 \end{pmatrix}.$$

Thus, $\mathcal{R}^1_0(A) = \mathbb{Z}_2^2$, $\mathcal{R}^1_1(A) = \{x_1x_2 + x_1 = 0\}$, and $\mathcal{R}^1_2(A) = \emptyset$.

3.3. Naturality properties. The next result describes a (partial) functoriality property enjoyed by the resonance varieties. Other results of this type (in various levels of generality, though not in the generality we work in here) can be found in [**18**] and [**35**].

PROPOSITION 3.5. *Let $\varphi\colon (A, \mathrm{d}_A) \to (B, \mathrm{d}_B)$ be a morphism of finite-type* CDGA*s. Suppose φ^i is an isomorphism for all $i \le q$ and a monomorphism for $i = q + 1$, for some $q \ge 0$.*

(1) *If $q = 0$, the map $\bar{\varphi}\colon \mathrm{MC}(A) \to \mathrm{MC}(B)$ is an embedding which sends $\mathcal{R}^1_s(A)$ into $\mathcal{R}^1_s(B)$ for all $s \ge 0$.*
(2) *If $q \ge 1$, the map $\bar{\varphi}\colon \mathrm{MC}(A) \to \mathrm{MC}(B)$ is an isomorphism which identifies $\mathcal{R}^i_s(A)$ with $\mathcal{R}^i_s(B)$ for all $i \le q$ and sends $\mathcal{R}^{q+1}_s(A)$ into $\mathcal{R}^{q+1}_s(B)$, for all $s \ge 0$.*

PROOF. (1) First assume that $q = 0$. By our hypothesis, the map φ^1 is injective, and thus $\bar{\varphi}$ is also injective. Let $a \in \mathrm{MC}(A)$. By Lemma 2.7, part (1), the chain map $\varphi_a\colon (A^\bullet, \delta^A_a) \to (B^\bullet, \delta^B_{\bar{\varphi}(a)})$ induces a monomorphism, $\varphi^1_a\colon H^1(A, \delta^A_a) \to H^1(B, \delta^B_{\bar{\varphi}(a)})$. Now suppose $a \in \mathcal{R}^1_s(A)$, that is, $\dim_{\Bbbk} H^1(A, \delta^A_a) \ge s$; then $H^1(B, \delta^B_{\bar{\varphi}(a)}) \ge s$, and so $\bar{\varphi}(a) \in \mathcal{R}^1_s(B)$. This shows that $\bar{\varphi}(\mathcal{R}^1_s(A)) \subseteq \mathcal{R}^1_s(B)$, and the first claim is proved.

(2) Next, assume that $q \ge 1$. By hypothesis, the map φ^1 is an isomorphism; thus, $\bar{\varphi}$ is also an isomorphism. For each $a \in \mathrm{MC}(A)$, Lemma 2.7 implies that the homomorphism $\varphi^i_a\colon H^i(A, \delta^A_a) \to H^i(B, \delta^B_{\bar{\varphi}(a)})$ is an isomorphism for $i \le q$ and a monomorphism for $i = q + 1$. The second claim follows. □

A similar argument yields the following proposition, which shows that the resonance varieties of a $\Bbbk$-CDGA (A, d) only depend on the isomorphism type of A.

PROPOSITION 3.6. *Let $\varphi\colon A \to B$ be an isomorphism of* CDGA*s. Then φ restricts to an isomorphism $\bar{\varphi}\colon \mathrm{MC}(A) \to \mathrm{MC}(B)$ which identifies $\mathcal{R}^q_s(A)$ with $\mathcal{R}^q_s(B)$ for all $q, s \ge 0$.*

The conclusions of Proposition 3.6 do not always hold if we only assume that the map $\varphi\colon A \to B$ is a quasi-isomorphism. We give a simple example of this phenomenon, adapted from [**18**] and [**34**].

EXAMPLE 3.7. Let $A = (\bigwedge(a,b), \mathrm{d})$ be the CDGA from Example 2.11, with $\mathrm{d}(a) = 0$ and $\mathrm{d}(b) = ba$, and let $A' = (\bigwedge(a), 0)$ be the sub-CDGA generated by a. It is readily seen that the inclusion map, $\varphi\colon A' \hookrightarrow A$, induces an isomorphism $\varphi^*\colon H^\bullet(A') \xrightarrow{\simeq} H^\bullet(A)$; moreover, φ^1 identifies $\mathrm{MC}(A') = \Bbbk$ with $\mathrm{MC}(A)$. On the other hand, $\mathcal{R}^1_1(A') = \{0\}$ is strictly included in $\mathcal{R}^1_1(A) = \{0, 1\}$.

Noteworthy is the fact that the resonance varieties are compatible with restriction and extension of the base field.

LEMMA 3.8. *Let A be a finite-type $\Bbbk$-CDGA, and let $\Bbbk \subset \mathbb{K}$ be a field extension. Then, for all $q, s \geq 0$,*

(1) $\mathcal{R}^q_s(A) = \mathcal{R}^q_s(A \otimes_\Bbbk \mathbb{K}) \cap \mathrm{MC}(A)$.
(2) $\mathcal{R}^q_s(A \otimes_\Bbbk \mathbb{K}) = \mathcal{R}^q_s(A) \times_\Bbbk \mathbb{K}$.

PROOF. By (24), the resonance varieties $\mathcal{R}^q_s(A)$ are determinantal varieties of matrices defined over $\Bbbk$. The two claims follow. □

3.4. Resonance varieties of tensor products. Let (A, d_A) and (B, d_B) be two $\Bbbk$-CDGAs. The tensor product of these two $\Bbbk$-vector spaces, $A \otimes_\Bbbk B$, has a natural CDGA structure, with grading $(A \otimes_\Bbbk B)^q = \bigoplus_{i+j=q} A^i \otimes_\Bbbk B^j$, multiplication $(a \otimes b) \cdot (a' \otimes b') = (-1)^{|b||a'|}(ab \otimes a'b')$, and differential D given on homogeneous elements by $D(a \otimes b) = d_A(a) \otimes b + (-1)^{|a|} a \otimes d_B(b)$.

If both A and B are connected, finite-type CDGAs then clearly $A \otimes_\Bbbk B$ is also a connected, finite-type CDGA. Moreover, upon identifying $(A \otimes_\Bbbk B)^1$ with $A^1 \oplus B^1 = A^1 \times B^1$, we have

$$\mathrm{MC}(A \otimes_\Bbbk B) = \mathrm{MC}(A) \times \mathrm{MC}(B). \tag{25}$$

The resonance varieties of a tensor product of CDGAs obey a type of Künneth formula. Such formulas were obtained in [**24**], [**38**], and [**35**] under the assumption that the differentials vanish; a proof of one inclusion was given in full generality in [**25**]. We give here a proof which is both complete and in full generality.

PROPOSITION 3.9. *Let (A, d_A) and (B, d_B) be two connected, finite-type* CDGAs. *Then,*

(1) $\mathcal{R}^q_1(A \otimes_\Bbbk B) = \bigcup_{i+j=q} \mathcal{R}^i_1(A) \times \mathcal{R}^j_1(B)$, *for all* $q \geq 1$.
(2) $\mathcal{R}^1_s(A \otimes_\Bbbk B) = \mathcal{R}^1_s(A) \times \{0\} \cup \{0\} \times \mathcal{R}^1_s(B)$, *for all* $s \geq 1$.

PROOF. Set $C = A \otimes_\Bbbk B$, and let $c = (a, b)$ be an element in $\mathrm{MC}(C) = \mathrm{MC}(A) \times \mathrm{MC}(B)$. The cochain complex (C, δ^C_c) splits as a tensor product of cochain complexes, $(A, \delta^A_a) \otimes_\Bbbk (B, \delta^B_b)$. Therefore,

$$b_q(C, c) = \sum_{i+j=q} b_i(A, a) \cdot b_j(B, b), \tag{26}$$

and part (1) follows. When $q = 1$, we obtain from (26) the following equalities:

$$\begin{aligned}
&b_1(C, (0,0)) = b_1(A, 0) + b_1(B, 0), && \\
&b_1(C, (a,0)) = b_1(A, a) && \text{if } a \neq 0, \\
&b_1(C, (0,b)) = b_1(B, b) && \text{if } b \neq 0, \\
&b_1(C, (a,b)) = 0 && \text{if } a \neq 0 \text{ and } b \neq 0.
\end{aligned} \tag{27}$$

Part (2) readily follows. □

3.5. Resonance varieties of coproducts. Let (A, d_A) and (B, d_B) be two connected CDGAs. Their wedge sum, $A \vee B$, is a new connected CDGA, whose underlying graded vector space in positive degrees is the direct sum $A^+ \oplus B^+$, whose multiplication is given by $(a, b) \cdot (a', b') = (aa', bb')$, and whose differential is $D = d_A + d_B$. Note that $(A \vee B)^1 = A^1 \oplus B^1$; therefore, we may identify the Maurer–Cartan set $\mathrm{MC}(A \vee B)$ with the product $\mathrm{MC}(A) \times \mathrm{MC}(B)$.

The following proposition recovers a result proved in various levels of generality in [**24**], [**25**], [**38**], and [**35**].

PROPOSITION 3.10. *Let $C = A \vee B$ be the coproduct of two connected, finite-type* CDGA*s with $b_1(A) > 0$ and $b_1(B) > 0$. Then, we have for all $s \geq 1$*

$$\mathcal{R}^q_s(A \vee B) = \begin{cases} \bigcup_{j+k=s-1} \mathcal{R}^1_j(A) \times \mathcal{R}^1_k(B) & \text{if } q = 1, \\ \bigcup_{j+k=s} \mathcal{R}^q_j(A) \times \mathcal{R}^q_k(B) & \text{if } q \geq 2. \end{cases} \tag{28}$$

PROOF. Let $c = (a, b)$ be an element in $\mathrm{MC}(C) = \mathrm{MC}(A) \times \mathrm{MC}(B)$. The chain complex (C, c) splits (in positive degrees) as a direct sum of chain complexes, $(C^+, \delta^C_c) \cong (A^+, \delta^A_a) \oplus (B^+, \delta^B_b)$. The Betti numbers of the respective chain complexes are related, as follows:

$$b_q(C, c) = \begin{cases} b_q(A, a) + b_q(B, b) + 1 & \text{if } q = 1 \text{ and } a \neq 0,\, b \neq 0, \\ b_q(A, a) + b_q(B, b) & \text{otherwise.} \end{cases} \tag{29}$$

The claim now follows by a case-by-case analysis of formula (29). □

4. Poincaré duality algebras and resonance varieties

In this section we study the interplay between Poincaré duality, the Koszul complex, and the resonance varieties of a differential graded algebra.

4.1. Poincaré duality. We start with a basic algebraic concept that abstracts the classical notion of Poincaré duality for manifolds. For more information on this subject, we refer to [**21**], [**30**], and [**35**].

DEFINITION 4.1. Let A be a connected, finite-type CGA over a field $\Bbbk$. We say that A is a *Poincaré duality $\Bbbk$-algebra* of formal dimension m (for short, an m-PDA) if there is a $\Bbbk$-linear map $\varepsilon\colon A^m \to \Bbbk$ (called an *orientation*) such that the bilinear forms

$$A^i \otimes_{\Bbbk} A^{m-i} \to \Bbbk, \quad a \otimes b \mapsto \varepsilon(ab) \tag{30}$$

are non-singular, for all $i \geq 0$.

It follows straight from the definition that the map ε is an isomorphism, and that $A^i = 0$ for $i > m$. Furthermore, for each $0 \leq i \leq m$, there is an isomorphism

$$\mathrm{PD}^i\colon A^i \to (A^{m-i})^\vee, \quad \mathrm{PD}^i(a)(b) = \varepsilon(ab). \tag{31}$$

Consequently, each element $a \in A^i$ has a Poincaré dual, $a^\vee \in A^{m-i}$, which is uniquely determined by the formula $\varepsilon(aa^\vee) = 1$. We define the orientation class $\omega_A \in A^m$ as the Poincaré dual of $1 \in A^0$, that is, $\omega_A = 1^\vee$. Conversely, a choice of orientation class $\omega_A \in A^m$ defines an orientation $\varepsilon\colon A^m \to \Bbbk$ by setting $\varepsilon(\omega_A) = 1$.

DEFINITION 4.2. A $\Bbbk$-CDGA $(A^\bullet, \mathrm{d})$ is a *Poincaré duality differential graded algebra* of formal dimension m (for short, an m-PD-CDGA) if

(1) The underlying graded algebra $A^{\bullet}$ is an m-PDA.
(2) $\mathrm{d}(A^{m-1}) = 0$.

Condition (2) can also be stated as $\varepsilon(\mathrm{d}(u)) = 0$ for all $u \in A^{m-1}$. By condition (1), the algebra A is connected and $A^m \cong A^0$; thus, condition (2) is equivalent to $H^m(A) = \Bbbk$. As noted in [**14**], if $(A^{\bullet}, \mathrm{d})$ is an m-PD-CDGA, then $H^{\bullet}(A)$ is an m-PDA, with orientation $[\varepsilon]\colon H^m(A) \to \Bbbk$ given by $[\varepsilon]([u]) = \varepsilon(u)$ for every $u \in Z^m(A)$. Lambrechts and Stanley also showed in [**15**] that if A is a CDGA such that $H^1(A) = 0$ and $H^{\bullet}(A)$ is an m-PDA, then A is weakly equivalent to an m-PD-CDGA.

Clearly, if A is an m-PDA, then A endowed with the zero differential is an m-PD-CDGA. Also, if (A, d) is a CDGA such that $A^1 = 0$ and A is an m-PDA, then $A^{m-1} = 0$, and so (A, d) is m-PD-CDGA. We give a simple example of a PD-CDGA (A, d) with $\mathrm{d} \neq 0$ and $A^1 \neq 0$.

EXAMPLE 4.3. Let $A = (\mathbb{Z}_2[x]/(x^{2k}), \mathrm{d})$, where $|x| = 1$, $k \geq 2$, and $\mathrm{d}(x) = x^2$. Clearly, A is a Poincaré duality algebra of dimension $m = 2k-1$; since $\mathrm{d}(x^{2k-2}) = 0$, we have that A is an m-PD-CDGA.

Next, we provide an example of a CDGA (A, d) such that A is an m-PDA, yet condition (2) from Definition 4.2 is not satisfied.

EXAMPLE 4.4. Let $A = (\mathbb{Z}_2[x, y]/(x^2, y^2), \mathrm{d})$ with $|x| = 1$, $|y| = k$, and differential given by $\mathrm{d}(x) = 0$, $\mathrm{d}(y) = xy$. Clearly, the underlying graded algebra $A^{\bullet}$ is a $\mathbb{Z}_2$-Poincaré duality algebra of dimension $k + 1$, with orientation class $\omega_A = xy$. Nevertheless, $\mathrm{d}(y) \neq 0$, and so (A, d) is not a PD-CDGA.

4.2. Resonance varieties of PD-cdgas. Poincaré duality imposes some rather stringent constraints on the resonance varieties of a PD-CDGA A. To ascertain those constraints, we start by analyzing the boundary maps in the Aomoto complexes (A, δ_a). Versions of the next lemma were given in [**27**, Lemma 7.4] when $\mathrm{char}(\Bbbk) = 0$ and in [**35**, Lemma 5.1] when $\mathrm{char}(\Bbbk) \neq 2$ and $\mathrm{d} = 0$. For completeness and to fix a sign issue in the latter reference, we give a self-contained proof here, valid in full generality.

LEMMA 4.5. *Let $(A^{\bullet}, \mathrm{d})$ be an m-PD-CDGA over an arbitrary field $\Bbbk$. Then, for all $a \in A^1$ and all $0 \leq i \leq m$, we have a commuting square,*

$$\begin{array}{ccc} (A^{m-i})^{\vee} & \xrightarrow{(\delta_{-a}^{m-i-1})^{\vee}} & (A^{m-i-1})^{\vee} \\ \cong \uparrow \mathrm{PD}^i & & \cong \uparrow (-1)^{i+1}\,\mathrm{PD}_{i+1} \\ A^i & \xrightarrow{\quad \delta_a^i \quad} & A^{i+1}\,. \end{array} \tag{32}$$

PROOF. Let $b \in A^i$ and $c \in A^{m-i-1}$. Then

$$\begin{aligned} (-1)^{i+1}\,\mathrm{PD}^{i+1} \circ \delta_a^i(b)(c) &= (-1)^{i+1}\,\mathrm{PD}^{i+1}(ab + \mathrm{d}(b))(c) \\ &= (-1)^{i+1}\varepsilon(abc + \mathrm{d}(b)c), \end{aligned} \tag{33}$$

while

$$\begin{aligned} (\delta_{-a}^{m-i-1})^{\vee} \circ \mathrm{PD}^i(b)(c) &= \mathrm{PD}^i(b)(\delta_{-a}^{m-i-1}(c)) \\ &= \mathrm{PD}^i(b)(-ac + \mathrm{d}(c)) \\ &= \varepsilon(-bac + b\,\mathrm{d}(c)). \end{aligned} \tag{34}$$

Now, since A is a CDGA, we have that $ab = (-1)^i ba$ and $\mathrm{d}(b)c + (-1)^i b\,\mathrm{d}(c) = \mathrm{d}(bc)$. Moreover, since A is an m-PD-CDGA and $bc \in A^{m-1}$, we have that $\mathrm{d}(bc) = 0$, and the claim follows. □

We are now in a position to state and prove the main result of this section, which recovers results from [**27**] and [**35**] and extends them to our present context. First recall from Lemma 2.4 that the Maurer–Cartan set $\mathrm{MC}(A)$ is invariant under the involution of A^1 sending a to $-a$.

THEOREM 4.6. *Let $(A^\bullet, \mathrm{d})$ be a Poincaré duality* CDGA *of formal dimension m over a field $\Bbbk$. Then,*

1. *$H^i(A, \delta_a)^\vee \cong H^{m-i}(A, \delta_{-a})$ for all $a \in \mathrm{MC}(A)$ and $i \geq 0$.*
2. *The linear isomorphism $A^1 \xrightarrow{\simeq} A^1$, $a \mapsto -a$ restricts to isomorphisms $\mathcal{R}^i_s(A) \xrightarrow{\simeq} \mathcal{R}^{m-i}_s(A)$ for all $i, s \geq 0$.*
3. *$\mathcal{R}^m_1(A) = \{0\}$.*

PROOF. Part (1) is a direct consequence of Lemma 4.5. Part (2) follows from (1). Finally, Part (3) follows from (2) and the fact that A is connected, and so $\mathcal{R}^0_1(A) = \{0\}$. □

5. The Aomoto–Bockstein complex

We now use the mod-2 cohomology algebra of a topological space X, equipped with the differential given by the Bockstein operator to define the Aomoto–Bockstein complex of X.

5.1. The Bockstein operator. We start by reviewing some standard material on Bocksteins, following the treatment in Hatcher [**11**, §3.E]. For each prime p, let $A = H^\bullet(X, \mathbb{Z}_p)$ be the cohomology algebra of X with $\mathbb{Z}_p$ coefficients, with multiplication given by the cup-product. For each $q \geq 0$, we have a *Bockstein operator*, $\beta_p\colon H^q(X, \mathbb{Z}_p) \to H^{q+1}(X, \mathbb{Z}_p)$, which is defined as the coboundary homomorphism associated to the coefficient exact sequence

$$0 \longrightarrow \mathbb{Z}_p \xrightarrow{\times p} \mathbb{Z}_{p^2} \longrightarrow \mathbb{Z}_p \longrightarrow 0. \tag{35}$$

Alternatively, if $\beta_0\colon H^q(X, \mathbb{Z}_p) \to H^{q+1}(X, \mathbb{Z})$ is the coboundary map associated to the coefficient sequence $0 \to \mathbb{Z} \to \mathbb{Z} \to \mathbb{Z}_p \to 0$ and $\rho_p\colon H^\bullet(X, \mathbb{Z}) \to H^\bullet(X, \mathbb{Z}_p)$ is the morphism induced by the projection $\mathbb{Z} \twoheadrightarrow \mathbb{Z}_p$, then $\beta_p = \rho_p \circ \beta_0$.

Note that a cohomology class $u \in H^q(X, \mathbb{Z}_p)$ is the reduction mod-p of a cohomology class in $H^q(X, \mathbb{Z}_{p^2})$ if and only if $\beta_p(u) = 0$. Consequently, if $u = \rho_p(v)$ for some $v \in H^q(X, \mathbb{Z})$, then $\beta_p(u) = 0$. Hence, if all the integral cohomology groups of X are torsion-free, the Bockstein β_p vanishes identically.

The Bockstein operators in all degrees $q \geq 0$ assemble into a $\mathbb{Z}_p$-linear map, $\beta_p\colon A \to A$, with the properties that $\beta_p \circ \beta_p = 0$ and $\beta_p(a \cup b) = \beta_p(a) \cup b + (-1)^{|a|} a \cup \beta_p(b)$ for all homogeneous elements $a, b \in A$. Therefore, β_p is a graded derivation of the algebra A and the pair (A, β_p) is a CDGA. Furthermore, the Bockstein is functorial: if $f\colon X \to Y$ is a continuous map, and $f^*\colon H^\bullet(Y, \mathbb{Z}_p) \to H^\bullet(X, \mathbb{Z}_p)$ is the induced homomorphism in cohomology, then $\beta^X_p(f^*(u)) = f^*(\beta^Y_p(u))$, for all $u \in H^\bullet(Y, \mathbb{Z}_p)$.

The homology groups of the Bockstein cochain complex (A, β_p), denoted $BH^q(X, \mathbb{Z}_p)$, are called the *Bockstein cohomology groups* of X (with coefficients in $\mathbb{Z}_p$). If $H^\bullet(X,\mathbb{Z})$ is of finite-type, then each $\mathbb{Z}$-summand of $H^q(X,\mathbb{Z})$ contributes a $\mathbb{Z}_p$-summand to $BH^q(X,\mathbb{Z}_p)$; the $\mathbb{Z}_p$-summands do not contribute to Bockstein cohomology; and each $\mathbb{Z}_{p^k}$-summand of $H^q(X,\mathbb{Z})$ with $k>1$ contributes a $\mathbb{Z}_p$-summand to both $BH^{q-1}(X,\mathbb{Z}_p)$ and $BH^q(X,\mathbb{Z}_p)$. In particular, $b_q(X) \le \dim_{\mathbb{Z}_p} BH^q(X,\mathbb{Z}_p)$.

When $p=2$, the Bockstein operator β_2 may be identified with the Steenrod square operation $\mathrm{Sq}^1\colon H^\bullet(X,\mathbb{Z}_2) \to H^\bullet(X,\mathbb{Z}_2)$. The fact that $\beta_2 = \mathrm{Sq}^1$ is a derivation may be seen as a consequence of the Ádem relation $\mathrm{Sq}^1 \circ \mathrm{Sq}^1 = 0$ and the Cartan relation $\mathrm{Sq}^1(a\cup b) = \mathrm{Sq}^1(a)\cup \mathrm{Sq}^0(b) + \mathrm{Sq}^0(a)\cup \mathrm{Sq}^1(b)$, where $\mathrm{Sq}^0 = \mathrm{id}$.

5.2. The Aomoto–Bockstein cochain complex. Assume now that X is a connected, finite-type CW-complex, and consider the cohomology algebra $A = H^\bullet(X,\mathbb{Z}_2)$, endowed with the differential given by the Bockstein operator $\beta_2 = \mathrm{Sq}^1$. Since $\mathrm{Sq}^1(a) = a^2$ for all $a\in A^1$, the Maurer–Cartan set for the CDGA (A,β_2) is then $\mathrm{MC}(A) = A^1$.

DEFINITION 5.1. The *Aomoto–Bockstein complex* of $A = H^\bullet(X,\mathbb{Z}_2)$ with respect to an element $a\in A^1$ is the cochain complex of finite-dimensional $\mathbb{Z}_2$-vector spaces,

$$(A,\delta_a)\colon\ A^0 \xrightarrow{\delta_a} A^1 \xrightarrow{\delta_a} \cdots \xrightarrow{\delta_a} A^q \xrightarrow{\delta_a} A^{q+1} \xrightarrow{\delta_a} \cdots, \tag{36}$$

where $\delta_a(u) = au + \beta_2(u)$.

Note that for $a=0$ this is the usual Bockstein cochain complex, (A,β_2); therefore, $H^q(A,\delta_0) = BH^q(X,\mathbb{Z}_2)$. On the other hand, if the Bockstein operator vanishes identically, then (36) is the usual Aomoto complex, with differentials $\delta_a(u) = au$.

Here is an alternative (and, in some ways, more convenient) interpretation. We pick a basis $\{e_1,\dots,e_n\}$ for $A^1 = H^1(X,\mathbb{Z}_2)$, we let $\{x_1,\dots,x_n\}$ be the Kronecker dual basis for $A_1 = H_1(X,\mathbb{Z}_2)$, and we identify the symmetric algebra $\mathrm{Sym}(A_1)$ with the polynomial ring $\mathbb{Z}_2[x_1,\dots,x_n]$. The coordinate ring of A^1 is then the quotient ring

$$S = \mathbb{Z}_2[x_1,\dots,x_n]/(x_1^2+x_1,\dots,x_n^2+x_n), \tag{37}$$

which may be viewed as the ring of (Boolean) functions on $\mathbb{Z}_2^n$. For detailed information on how to compute with polynomials, ideals, Gröbner bases, and varieties over this ring, we refer to [**1**], [**2**], and [**17**].

DEFINITION 5.2. The *universal Aomoto–Bockstein complex* of the cohomology algebra $A = H^\bullet(X,\mathbb{Z}_2)$ is the cochain complex of free S-modules,

$$\begin{aligned}(A\otimes_{\mathbb{Z}_2} S,\delta)\colon\ & A^0\otimes_{\mathbb{Z}_2} S \xrightarrow{\delta^0} A^1\otimes_{\mathbb{Z}_2} S \longrightarrow \cdots \\ & \longrightarrow A^i\otimes_{\mathbb{Z}_2} S \xrightarrow{\delta^i} A^{i+1}\otimes_{\mathbb{Z}_2} S \longrightarrow \cdots,\end{aligned} \tag{38}$$

where the differentials are defined by

$$\delta^i(u\otimes 1) = \sum_{j=1}^n e_j u\otimes x_j + \beta_2(u)\otimes 1 \tag{39}$$

for $u\in A^i$, and then extended by S-linearity.

If the Bockstein operator acts trivially on $H^{\bullet}(X, \mathbb{Z}_2)$, then each boundary map δ^i in the cochain complex (38) is given by a matrix whose entries are linear forms in the variables $x_1, \dots, x_n$. In general, though, the entries of δ^i may also have non-zero constant terms.

EXAMPLE 5.3. Let $X = \mathbb{RP}^{\infty}$ be the infinite-dimensional real projective space. We identify $H^{\bullet}(X, \mathbb{Z}_2) = \mathbb{Z}_2[a]$ where $|a| = 1$ and note that the Bockstein β_2 sends a^i to a^{i+1} if i is odd and to 0 otherwise. Writing $S = \mathbb{Z}_2[x]/(x^2 + x)$, the resulting Aomoto–Bockstein complex has boundary maps $\delta^i \colon S \to S$ given by $\delta^i(a^i) = (x+1)a^{i+1}$ if i is odd and $\delta^i(a^i) = xa^{i+1}$ if i is even, and thus has the form

$$S \xrightarrow{\ x\ } S \xrightarrow{\ x+1\ } S \xrightarrow{\ x\ } S \longrightarrow \cdots . \tag{40}$$

Note that this cochain complex is exact.

5.3. The twisted Bockstein operator. To conclude this section, we present an alternative interpretation of the differentials in the Aomoto–Bockstein complex (36), due to Samelson [**29**] and Greenblatt [**10**].

Fix a basepoint $x_0 \in X$ and let $\pi = \pi_1(X, x_0)$ be the fundamental group of X at x_0. Every mod 2 cohomology class $w \in H^1(X, \mathbb{Z}_2)$ defines a local system $\mathbb{Z}_w$ on X, as follows. Viewing w as a homomorphism from $H_1(X, \mathbb{Z})$ to $\mathbb{Z}_2$, and identifying the target group with $\{\pm 1\}$, regarded as the units of the ring $\mathbb{Z}$ puts a $\mathbb{Z}[H_1(X, \mathbb{Z})]$-module structure on the additive group $\mathbb{Z}$. Restricting scalars to the group ring $\mathbb{Z}[\pi]$ via the abelianization map $\pi \twoheadrightarrow \pi_{\mathrm{ab}} = H_1(X, \mathbb{Z})$ yields a $\mathbb{Z}[\pi]$-module structure on $\mathbb{Z}$, which is the module that we denote by $\mathbb{Z}_w$.

The twisted Bockstein operator $\beta_w \colon H^q(X, \mathbb{Z}_2) \to H^{q+1}(X, \mathbb{Z}_w)$ is defined as the coboundary map induced by the exact sequence

$$0 \longrightarrow \mathbb{Z}_w \xrightarrow{\ \times 2\ } \mathbb{Z}_w \longrightarrow \mathbb{Z}_2 \longrightarrow 0. \tag{41}$$

Let $\bar{\rho}_2 \colon H^{\bullet}(X, \mathbb{Z}_w) \to H^{\bullet}(X, \mathbb{Z}_2)$ be the coefficient morphism induced by the projection $\mathbb{Z}_w \twoheadrightarrow \mathbb{Z}_2$, and set $\bar{\beta}_w := \bar{\rho}_2 \circ \beta_w \colon H^q(X, \mathbb{Z}_2) \to H^{q+1}(X, \mathbb{Z}_2)$. By standard properties of the Bockstein operator, we have that $\bar{\beta}_w^2 = 0$. Furthermore, as shown in [**10**], the map

$$\bar{\beta} \colon H^1(X, \mathbb{Z}_2) \times H^q(X, \mathbb{Z}_2) \longrightarrow H^{q+1}(X, \mathbb{Z}_2) \tag{42}$$

given by $\bar{\beta}(w, u) = \bar{\beta}_w(u)$ is a cohomology operation. That is, if $f \colon X \to Y$ is a continuous map, and $f^* \colon H^{\bullet}(Y, \mathbb{Z}_2) \to H^{\bullet}(X, \mathbb{Z}_2)$ is the induced homomorphism in cohomology, then, for every $w \in H^1(X, \mathbb{Z}_2)$ and $u \in H^q(Y, \mathbb{Z}_2)$, we have that

$$f^*(\bar{\beta}(w, u)) = \bar{\beta}(f^*(w), f^*(u)). \tag{43}$$

THEOREM 5.4 ([**29**] and [**10**]). *For every $w \in H^1(X, \mathbb{Z}_2)$ and $u \in H^q(X, \mathbb{Z}_2)$, we have that $\bar{\beta}(w, u) = wu + \mathrm{Sq}^1(u)$.*

6. The resonance varieties of a space

In this section we discuss the resonance varieties of a space, with coefficients in a field $\Bbbk$. Throughout, X will be a connected, finite-type CW-complex and $A = H^{\bullet}(X, \Bbbk)$ will be its cohomology algebra over $\Bbbk$, with multiplication given by the cup-product.

6.1. The usual resonance varieties. We start by singling out certain situations in which all cohomology classes in degree 1 square to zero. The next (standard) lemma formalizes arguments from [**23**] and [**24**].

LEMMA 6.1. *Suppose either* $\operatorname{char}(\Bbbk) \neq 2$, *or* $\operatorname{char}(\Bbbk) = 2$ *and* $H_1(X, \mathbb{Z})$ *has no 2-torsion. Then* $a^2 = 0$ *for all* $a \in H^1(X, \Bbbk)$.

PROOF. By graded-commutativity of the cup-product, $a^2 = -a^2$. Thus, if $\operatorname{char}(\Bbbk) \neq 2$, then $a^2 = 0$.

Now suppose $\operatorname{char}(\Bbbk) = 2$ and $H_1(X, \mathbb{Z})$ has no 2-torsion. In this case, by the Universal Coefficients Theorem, every class $a \in H^1(X, \Bbbk)$ is the image of a class $\tilde{a} \in H^1(X, \mathbb{Z})$ under the coefficient homomorphism $\mathbb{Z} \to \Bbbk$. By obstruction theory, there is a map $f\colon X \to S^1$ and a class $\omega \in H^1(S^1, \mathbb{Z})$ such that $\tilde{a} = f^*(\omega)$. Hence, $\tilde{a} \cup \tilde{a} = f^*(\omega \cup \omega) = 0$. The claim then follows by naturality of cup products with respect to coefficient homomorphisms. □

Under the assumptions of Lemma 6.1, we shall view the cohomology algebra $H^\bullet(X, \Bbbk)$ as a $\Bbbk$-cdga with differential $d = 0$ and we shall identify the Maurer–Cartan set $\operatorname{MC}(A)$ with the affine space $A^1 = H^1(X, \Bbbk)$.

DEFINITION 6.2. Let X be a connected, finite-type CW-complex, let $\Bbbk$ be a field, and suppose either $\operatorname{char}(\Bbbk) \neq 2$, or $\operatorname{char}(\Bbbk) = 2$ and $H_1(X, \mathbb{Z})$ has no 2-torsion. The *resonance varieties* of X (over $\Bbbk$) are the resonance varieties of the CDGA $A = (H^*(X, \Bbbk), 0)$; that is,

$$\mathcal{R}^q_s(X, \Bbbk) = \{a \in H^1(X, \Bbbk) \mid \dim_{\Bbbk} H^q(A, \delta_a) \geq s\}, \tag{44}$$

where $\delta_a\colon A^q \to A^{q+1}$ is given by $\delta_a(u) = au$.

These sets are homogeneous algebraic subvarieties of the affine space $H^1(X, \Bbbk)$. For each $q \geq 0$, they form a descending filtration,

$$H^1(X, \Bbbk) = \mathcal{R}^q_0(X, \Bbbk) \supseteq \mathcal{R}^q_1(X, \Bbbk) \supseteq \mathcal{R}^q_2(X, \Bbbk) \supseteq \cdots \supseteq \mathcal{R}^q_{b_q}(X, \Bbbk) \supseteq \{0\}, \tag{45}$$

where $b_q = b_q(X, \Bbbk) := \dim_{\Bbbk} H^q(X, \Bbbk)$; moreover, $\mathcal{R}^q_{b_q+1}(X, \Bbbk) = \emptyset$. By construction, the resonance varieties of X depend only on the cohomology algebra $H^\bullet(X, \Bbbk)$. We illustrate the concept with some well-known examples, see e.g. [**32**] and [**34**].

EXAMPLE 6.3. Let T^n be the n-dimensional torus. Using the exactness of the Koszul complex from Example 2.10, we see that $\mathcal{R}^i_s(T^n, \Bbbk)$ is equal to $\{0\}$ if $0 < s \leq \binom{n}{i}$ and is empty for $s > \binom{n}{i}$.

EXAMPLE 6.4. Let M_g be the orientable surface of genus $g > 1$. In depth $s > 0$ the resonance varieties are then given by

$$\mathcal{R}^q_s(M_g, \Bbbk) = \begin{cases} \Bbbk^{2g} & \text{if } q = 1,\ s < 2g-1, \\ \{0\} & \text{if } q = 1,\ s \in \{2g-1, 2g\} \text{ or } q \in \{0, 2\},\ s = 1, \\ \emptyset & \text{otherwise.} \end{cases} \tag{46}$$

Best understood are the degree 1, depth 1 resonance varieties $\mathcal{R}^1_1(X, \Bbbk)$. For low values of $n = b_1(X, \Bbbk)$, these varieties are easy to describe. If $n = 0$, then $\mathcal{R}^1_1(X, \Bbbk) = \emptyset$, while if $n = 1$, then $\mathcal{R}^1_1(X, \Bbbk) = \{0\}$. If $n = 2$, then $\mathcal{R}^1_1(X, \Bbbk) = \Bbbk^2$ or $\{0\}$, according to whether the cup product vanishes on $H^1(X, \Bbbk)$ or not. For $n \geq 3$, the variety $\mathcal{R}^1_1(X, \Bbbk)$ can be much more complicated; in particular, it may have irreducible components which are not linear subspaces.

EXAMPLE 6.5. Let $X = \mathrm{Conf}(E,3)$ be the configuration space of 3 labeled points on a torus. Then $H^\bullet(X,\Bbbk)$ is the exterior algebra on generators $a_1, a_2, a_3, b_1, b_2, b_3$ in degree 1, modulo the ideal generated by$(a_1-a_2)(b_1-b_2)$, $(a_1-a_3)(b_1-b_3)$, and $(a_2-a_3)(b_2-b_3)$. A calculation reveals that

$$\mathcal{R}^1_1(X,\Bbbk) = \{(a,b)\in \Bbbk^6 \mid a_1+a_2+a_3 = b_1+b_2+b_3 = a_1b_2 - a_2b_1 = 0\}.$$

Hence, $\mathcal{R}^1_1(X,\Bbbk)$ is isomorphic to $\{a_1b_2-a_2b_1=0\}$, a smooth quadric hypersurface in $\Bbbk^4$.

6.2. Resonance varieties in characteristic 2. We now treat the case when $\mathrm{char}(\Bbbk)=2$, without imposing any restriction on the 2-torsion of $H_1(X,\mathbb{Z})$.

DEFINITION 6.6. The *Bockstein resonance varieties* of a connected, finite-type CW-complex X are the resonance varieties of the CDGA $A = (H^\bullet(X,\mathbb{Z}_2),\beta_2)$; that is,

$$\widetilde{\mathcal{R}}^q_s(X,\mathbb{Z}_2) = \big\{a\in H^1(X,\mathbb{Z}_2) \mid \dim_{\mathbb{Z}_2} H^q(A,\delta_a)\geq s\big\}, \tag{47}$$

where $\beta_2 = \mathrm{Sq}^1\colon A^q\to A^{q+1}$ is the Bockstein operator and $\delta_a\colon A^q\to A^{q+1}$ is given by $\delta_a(u) = au+\beta_2(u)$.

More generally, we define the Bockstein resonance varieties of X over a field $\Bbbk$ of characteristic 2 as

$$\widetilde{\mathcal{R}}^q_s(X,\Bbbk) = \widetilde{\mathcal{R}}^q_s(X,\mathbb{Z}_2)\times_{\mathbb{Z}_2}\Bbbk\,. \tag{48}$$

These sets are algebraic subvarieties of the affine space $H^1(X,\Bbbk)$. For each $q\geq 0$, they form a descending filtration,

$$H^1(X,\Bbbk) = \widetilde{\mathcal{R}}^q_0(X,\Bbbk)\supseteq \widetilde{\mathcal{R}}^q_1(X,\Bbbk)\supseteq \widetilde{\mathcal{R}}^q_2(X,\Bbbk)\supseteq\cdots. \tag{49}$$

Since all the essential features of the resonance varieties in characteristic 2 already appear over the field with 2 elements, we will concentrate mainly on the sets $\widetilde{\mathcal{R}}^q_s(X,\mathbb{Z}_2)$.

6.3. Properties of the Bockstein resonance varieties. As a direct consequence of Lemma 3.3, we have the following result.

LEMMA 6.7. *Let X be a connected, finite-type CW-complex. Then,*

(1) *The element $0\in H^1(X,\mathbb{Z}_2)$ belongs to $\widetilde{\mathcal{R}}^q_s(X,\mathbb{Z}_2)$ if and only if*
$$\dim_{\mathbb{Z}_2} BH^q(X,\mathbb{Z}_2)\geq s.$$
(2) *$\widetilde{\mathcal{R}}^0_1(X,\mathbb{Z}_2)=\{0\}$ and $\widetilde{\mathcal{R}}^0_s(X,\mathbb{Z}_2)=\emptyset$ for $s>1$.*

Since $\mathrm{Sq}^1(u) = u^2$ for all $u\in H^1(X,\mathbb{Z}_2)$, the varieties $\widetilde{\mathcal{R}}^1_s(X,\mathbb{Z}_2)$ depend only on the truncated cohomology ring $H^{\leq 2}(X,\mathbb{Z}_2)$. In depth $s=1$, we can be even more concrete.

LEMMA 6.8. *Let X be a connected, finite-type CW-complex. Then,*

(1) *$0\in\widetilde{\mathcal{R}}^1_1(X,\mathbb{Z}_2)$ if and only if there is an element $u\in H^1(X,\mathbb{Z}_2)$ such that $u^2=0$.*
(2) *A non-zero element $a\in H^1(X,\mathbb{Z}_2)$ belongs to $\widetilde{\mathcal{R}}^1_1(X,\mathbb{Z}_2)$ if and only if there is an element $u\in H^1(X,\mathbb{Z}_2)$ such that $au+u^2=0$ and u is not proportional to a.*

For instance, if $H^1(X,\mathbb{Z}_2)=\{0,a\}$, then $\widetilde{\mathcal{R}}^1_1(X,\mathbb{Z}_2)$ is either equal to $\{0\}$ or is empty, according to whether $a^2=0$ or not.

It follows from Proposition 3.5, part (1) that both types of degree 1 resonance varieties, $\mathcal{R}^1_s(X,\Bbbk)$ and $\widetilde{\mathcal{R}}^1_s(X,\Bbbk)$, enjoy a (partial) naturality property.

PROPOSITION 6.9. *Let $f\colon X\to Y$ be a continuous map, let $\Bbbk$ be a field, and suppose that the induced homomorphism $f^*\colon H^1(Y,\Bbbk)\to H^1(X,\Bbbk)$ is injective.*

(1) *If either* $\operatorname{char}(\Bbbk)\neq 2$, *or* $\operatorname{char}(\Bbbk)=2$ *and both $H_1(X,\mathbb{Z})$ and $H_1(Y,\mathbb{Z})$ have no 2-torsion, then f^* restricts to embeddings $\mathcal{R}^1_s(Y,\Bbbk)\hookrightarrow\mathcal{R}^1_s(X,\Bbbk)$ for all $s\geq 1$.*

(2) *If* $\operatorname{char}(\Bbbk)=2$, *then f^* restricts to embeddings $\widetilde{\mathcal{R}}^1_s(Y,\Bbbk)\hookrightarrow\widetilde{\mathcal{R}}^1_s(X,\Bbbk)$ for all $s\geq 1$.*

6.4. Sample computations of resonance. We now give several examples illustrating the computation of the Bockstein resonance varieties $\widetilde{\mathcal{R}}^q_s(X,\mathbb{Z}_2)$ for some familiar spaces X. For simplicity, we will assume throughout that $s>0$.

EXAMPLE 6.10. Let N_g be the non-orientable surface of genus $g\geq 1$. Then $H^\bullet(N_g,\mathbb{Z}_2)=\mathbb{Z}_2[a_1,\dots,a_g]/I$, where $|a_i|=1$ and I is the ideal generated by $a_i^2+a_j^2$ and a_ia_j for all $1\leq i<j\leq g$. Direct computation shows that

$$\widetilde{\mathcal{R}}^q_s(N_g,\mathbb{Z}_2)=\begin{cases}\{0,a_1+\cdots+a_g\} & \text{if } q=1 \text{ and } s=1,\\ \{0\} & \text{if } q=0 \text{ or } 2 \text{ and } s=1,\\ \emptyset & \text{otherwise.}\end{cases}\tag{50}$$

EXAMPLE 6.11. Let $X=\mathbb{RP}^n$ be the real projective space of dimension n. Then $H^\bullet(\mathbb{RP}^n,\mathbb{Z}_2)=\mathbb{Z}_2[a]/(a^{n+1})$, with a in degree 1, and $\beta_2(a^q)=a^{q+1}$ if q is odd, and 0 otherwise. Thus, $BH^q(\mathbb{RP}^n,\mathbb{Z}_2)=0$ for $0<q<n$; moreover, $\delta_a(a^q)=a^{q+1}$ if a is even, and 0 otherwise. It follows that

$$\widetilde{\mathcal{R}}^q_s(\mathbb{RP}^n,\mathbb{Z}_2)=\begin{cases}\mathbb{Z}_2 & \text{if } q=n,\ n \text{ is even, and } s=1,\\ \{0\} & \text{if } q=0 \text{ and } s=1,\\ \emptyset & \text{otherwise.}\end{cases}\tag{51}$$

EXAMPLE 6.12. Let $\mathrm{Gr}_n=\mathrm{Gr}_n(\mathbb{R}^\infty)$ be the Grassmannian of (unoriented) n-planes in $\mathbb{R}^\infty$. Then $H^\bullet(\mathrm{Gr}_n,\mathbb{Z}_2)=\mathbb{Z}_2[w_1,\dots,w_n]$, where w_k is the (universal) Stiefel–Whitney class of degree k, and the Bockstein operator $\beta_2=\mathrm{Sq}^1$ is given by Wu's formula,

$$\beta_2(w_k)=w_1w_k+(k+1)w_{k+1},\tag{52}$$

see [**22**]. A straightforward computation (or the observation at the end of Example 5.3) shows that $\widetilde{\mathcal{R}}^q_s(\mathrm{Gr}_1,\mathbb{Z}_2)=\emptyset$ for all $q,s>0$. A slightly more involved computation yields

$$\widetilde{\mathcal{R}}^{2q}_s(\mathrm{Gr}_2,\mathbb{Z}_2)=\widetilde{\mathcal{R}}^{2q+1}_s(\mathrm{Gr}_2,\mathbb{Z}_2)=\begin{cases}\mathbb{Z}_2 & \text{if } 0<s\leq q,\\ \{0\} & \text{if } q<s\leq 2q,\\ \emptyset & \text{if } s>2q.\end{cases}\tag{53}$$

The resonance varieties $\widetilde{\mathcal{R}}^q_s(\mathrm{Gr}_n,\mathbb{Z}_2)$ for $n>2$ can be computed in like fashion.

6.5. Comparing the two types of resonance. We end this section with a comparison between the two types of mod-2 resonance varieties, $\mathcal{R}^q_s(X,\mathbb{Z}_2)$ and $\widetilde{\mathcal{R}}^q_s(X,\mathbb{Z}_2)$. We start by observing that these varieties do agree in degree $q=1$, provided the former are defined.

LEMMA 6.13. *Suppose $H_1(X,\mathbb{Z})$ has no 2-torsion. Then $\mathcal{R}^1_s(X,\mathbb{Z}_2)=\widetilde{\mathcal{R}}^1_s(X,\mathbb{Z}_2)$ for all $s\geq 0$.*

PROOF. By Lemma 6.1, we have that $u^2=0$ for all $u\in H^1(X,\mathbb{Z}_2)$. Thus, the Bockstein $\beta_2\colon H^1(X,\mathbb{Z}_2)\to H^2(X,\mathbb{Z}_2)$ vanishes, and the claim follows straight from the definitions of the respective resonance varieties. □

In degrees $q>1$, though, it may happen that $\mathcal{R}^q_s(X,\mathbb{Z}_2)\neq\widetilde{\mathcal{R}}^q_s(X,\mathbb{Z}_2)$. We shall see examples of spaces X for which $\mathcal{R}^q_1(X,\mathbb{Z}_2)\subsetneqq\widetilde{\mathcal{R}}^q_1(X,\mathbb{Z}_2)$ in some degrees $q>1$ in Example 6.15 below, and also in Corollary 7.11 and in Example 7.12 in the next section. We give now an example of a space X for which $\widetilde{\mathcal{R}}^q_1(X,\mathbb{Z}_2)\subsetneqq\mathcal{R}^q_1(X,\mathbb{Z}_2)$ for some $q>1$.

EXAMPLE 6.14. Let $X=S^1\vee\Sigma(\mathbb{RP}^2)$ be the wedge of a circle with a suspended projective plane. Then $H^i(X,\mathbb{Z}_2)=\mathbb{Z}_2$, generated by classes a_i for $1\leq i\leq 3$, with all cup products among these generators vanishing; moreover, since $\beta_2=\operatorname{Sq}^1$ commutes with suspensions, $\beta_2(a_2)=a_3$. It follows that $\mathcal{R}^2_1(X,\mathbb{Z}_2)=\mathbb{Z}_2$, whereas $\widetilde{\mathcal{R}}^2_1(X,\mathbb{Z}_2)=\emptyset$.

If the mod-2 cohomology ring of X is generated in degree 1, then the Bockstein $\beta_2=\operatorname{Sq}^1$ is completely determined by the cup-product structure of $H^\bullet(X,\mathbb{Z}_2)$, and so the mod-2 resonance varieties $\widetilde{\mathcal{R}}^q_s(X,\mathbb{Z}_2)$ depend only on the cohomology ring $H^\bullet(X,\mathbb{Z}_2)$. In general, though, the varieties $\widetilde{\mathcal{R}}^q_s(X,\mathbb{Z}_2)$ also depend on the action of the Steenrod algebra on the cohomology ring, as the next example shows.

EXAMPLE 6.15. Let $X=S^2\times S^1$ and let $Y=S^2\widetilde{\times}S^1$ be the non-trivial S^2-bundle over S^1, with monodromy given by the antipodal map. Clearly, $H_1(X,\mathbb{Z})=H_2(Y,\mathbb{Z})=\mathbb{Z}$. The mod-2 cohomology rings of both spaces are isomorphic to $\mathbb{Z}_2[a,b]/(a^2,b^2)$, where $|a|=1$ and $|b|=2$, while the two Bocksteins are given by $\beta^X_2(a)=\beta^X_2(b)=0$ and $\beta^Y_2(a)=0$, $\beta^Y_2(b)=ab$. The usual resonance varieties $\mathcal{R}^q_s$ of the two spaces are the same (over any field $\Bbbk$); in particular, $\mathcal{R}^2_1=\mathcal{R}^3_1=\{0\}$. Nevertheless, the mod-2 resonance varieties $\widetilde{\mathcal{R}}^q_s$ differ:

$$\begin{aligned}
&\widetilde{\mathcal{R}}^1_1(X,\mathbb{Z}_2)=\{0\}, &&\widetilde{\mathcal{R}}^2_1(X,\mathbb{Z}_2)=\{0\}, &&\widetilde{\mathcal{R}}^3_1(X,\mathbb{Z}_2)=\{0\},\\
&\widetilde{\mathcal{R}}^1_1(Y,\mathbb{Z}_2)=\{0\}, &&\widetilde{\mathcal{R}}^2_1(Y,\mathbb{Z}_2)=\mathbb{Z}_2, &&\widetilde{\mathcal{R}}^3_1(Y,\mathbb{Z}_2)=\mathbb{Z}_2.
\end{aligned}$$

7. Manifolds, Poincaré duality, and resonance

In this section we study the resonance varieties of topological manifolds. All manifolds will be assumed to be without boundary, compact, and connected—for short, *closed* manifolds. Most of the theory works as well for Poincaré complexes, though at some point we will require manifolds to be smooth. We start by discussing the interplay between Poincaré duality, orientability, and the Bockstein operator.

7.1. Poincaré duality and orientability. Let M be a closed manifold of dimension m. If M is orientable, then its cohomology algebra with coefficients in an arbitrary field $\Bbbk$ is a Poincaré duality algebra, also of dimension m. We consider here the case when M is not necessarily orientable, and restrict our attention to the coefficient field $\Bbbk = \mathbb{Z}_2$, in which case $H^\bullet(M, \mathbb{Z}_2)$ is again a Poincaré duality algebra of dimension m.

We start with a well-known lemma, relating the orientability of M to the Bockstein operator $\beta_2 = \mathrm{Sq}^1$ on $H^\bullet(M, \mathbb{Z}_2)$, see Massey [**19**, Lemma 1]. For completeness, we give a full proof, since only a brief sketch is given in that reference.

LEMMA 7.1 ([**19**]). *Let M be a closed manifold of dimension m. Then M is orientable if and only if the Bockstein $\beta_2\colon H^{m-1}(M, \mathbb{Z}_2) \to H^m(M, \mathbb{Z}_2)$ is zero.*

PROOF. If M is orientable, then $H_{m-1}(M, \mathbb{Z}) \cong H^1(M, \mathbb{Z})$ is free abelian. Thus, the Bockstein operator $\beta_0\colon H_m(M, \mathbb{Z}_2) \to H_{m-1}(M, \mathbb{Z})$ associated to the coefficient sequence $0 \to \mathbb{Z} \xrightarrow{\times 2} \mathbb{Z} \to \mathbb{Z}_2 \to 0$ is trivial. Since β_2 is $\mathbb{Z}_2$-dual to the homomorphism $\rho_2 \circ \beta_0\colon H_m(M, \mathbb{Z}_2) \to H_{m-1}(M, \mathbb{Z}_2)$, we infer that $\beta_2 = 0$.

If M is non-orientable, then the torsion subgroup of $H_{m-1}(M, \mathbb{Z})$ is equal to $\mathbb{Z}_2$ and may be identified with the image of the Bockstein $\beta_0\colon H_m(M, \mathbb{Z}_2) \to H_{m-1}(M, \mathbb{Z})$, see [**11**, p. 238]. Since $\beta_2 = (\rho_2 \circ \beta_0)^*$, we conclude that $\beta_2 \neq 0$. □

REMARK 7.2. For a smooth m-manifold M, an alternative proof can be given, using the Wu formulas (see also [**12**, Corollary 9.8.5]). These formulas relate the Wu classes $v_i \in H^i(M, \mathbb{Z}_2)$, the Stiefel–Whitney classes $w_i \in H^i(M, \mathbb{Z}_2)$, and the Steenrod squares $\mathrm{Sq}^i\colon H^q(M, \mathbb{Z}_2) \to H^{q+i}(M, \mathbb{Z}_2)$ via

$$w = \mathrm{Sq}(v) \quad \text{and} \quad v \cup u = \mathrm{Sq}(u) \tag{54}$$

for all $u \in H^\bullet(M, \mathbb{Z}_2)$, where $w = \sum_{i \geq 0} w_i$ is the total Stiefel–Whitney class and likewise for v and Sq, see [**22**]. Interpreting these formulas in degree 1, we see that $w_1 = v_1$ and v_1 is Poincaré dual to $\varepsilon \circ \mathrm{Sq}^1 \in H^{m-1}(M, \mathbb{Z}_2)^*$. Since $\beta_2 = \mathrm{Sq}^1$ and since M is orientable if and only if $w_1 = 0$, the desired conclusion follows.

We now consider the graded $\mathbb{Z}_2$-algebra $A = H^\bullet(M, \mathbb{Z}_2)$, viewed as a CDGA with differential given by the Bockstein $\beta_2\colon A \to A$.

COROLLARY 7.3. *Let M be a closed manifold of dimension m. Then*

$$(H^\bullet(M, \mathbb{Z}_2), \beta_2)$$

is a Poincaré duality differential graded algebra (of dimension m) if and only if M is orientable.

PROOF. By Poincaré duality, the cohomology algebra $H^\bullet(M, \mathbb{Z}_2)$ is an m-PDA. The claim follows from Definition 4.2 and Lemma 7.1. □

REMARK 7.4. As shown by Postnikov in [**28**], every 3-dimensional Poincaré duality algebra over $\mathbb{Z}_2$ can be realized as the cohomology ring, $H^\bullet(M, \mathbb{Z}_2)$, of some closed (not necessarily orientable) 3-manifold M.

7.2. Resonance varieties of closed manifolds. Poincaré duality has strong implications on the nature of the resonance varieties of closed manifolds (and Poincaré complexes). First, as an immediate consequence of Theorem 4.6, we have the following result.

PROPOSITION 7.5. *Let M be a closed, orientable manifold of dimension m, and let $\Bbbk$ be a field of characteristic different from 2. Then $\mathcal{R}^q_s(M;\Bbbk) = \mathcal{R}^{m-q}_s(M;\Bbbk)$ for all $q, s \geq 0$. In particular, $\mathcal{R}^m_1(M,\Bbbk) = \{0\}$.*

Turning to the resonance varieties over a field $\Bbbk$ of characteristic 2, we have several results. For the first one, we impose no orientability condition (over $\mathbb{Z}$), but make instead an assumption guaranteeing that the usual resonance varieties $\mathcal{R}^m_1(M,\Bbbk)$ are defined.

PROPOSITION 7.6. *Let M be a closed manifold of dimension m. Suppose $H_1(M,\mathbb{Z})$ has no 2-torsion and* $\operatorname{char}(\Bbbk) = 2$. *Then $\mathcal{R}^m_1(M,\Bbbk) = \{0\}$.*

PROOF. By Poincaré duality, the cohomology algebra $H^\bullet(M,\mathbb{Z}_2)$ is an m-PDA. Viewing it as an m-PD-CDGA with $\mathrm{d} = 0$, Theorem 4.6, Part (3) implies that $\mathcal{R}^m_1(M,\mathbb{Z}_2) = \{0\}$, and so $\mathcal{R}^m_1(M,\Bbbk) = \{0\}$, too.

The claim can also be proven directly, as follows. Let $\omega = 1^\vee$ be the generator of $H^m(M,\mathbb{Z}_2) = \mathbb{Z}_2$. Let $a \in H^1(M,\mathbb{Z}_2)$ and let $a^\vee \in H^{m-1}(M,\mathbb{Z}_2)$ be its Poincaré dual. Then $\delta_a(a^\vee) = aa^\vee = \omega$, and so $H^m(A,\delta_a) = 0$, once again showing that $\mathcal{R}^m_1(M,\mathbb{Z}_2) = \{0\}$. ☐

For the next result, we impose orientability (over $\mathbb{Z}$) but no other additional conditions on M, and draw some conclusions regarding the resonance varieties $\widetilde{\mathcal{R}}^m_1(M,\Bbbk)$.

PROPOSITION 7.7. *Let M be a closed, orientable manifold of dimension m, and assume* $\operatorname{char}(\Bbbk) = 2$. *Then $\widetilde{\mathcal{R}}^q_s(M;\Bbbk) = \widetilde{\mathcal{R}}^{m-q}_s(M;\Bbbk)$ for all $q, s \geq 0$. In particular, $\widetilde{\mathcal{R}}^m_1(M,\Bbbk) = \{0\}$.*

PROOF. By Corollary 7.3, the CDGA $\big(H^\bullet(M,\mathbb{Z}_2),\beta_2\big)$ is an m-PD-CDGA. Theorem 4.6 then gives $\widetilde{\mathcal{R}}^q_s(M;\mathbb{Z}_2) = \widetilde{\mathcal{R}}^{m-q}_s(M;\mathbb{Z}_2)$ for all $q, s \geq 0$, and the conclusions follow. ☐

EXAMPLE 7.8. The lens space $L(4,1) = S^3/\mathbb{Z}_4$ and the projective space $\mathbb{RP}^3 = S^3/\mathbb{Z}_2$ are both closed, orientable manifolds of dimension 3. They share the same $\mathbb{Z}_2$-cohomology groups, but the multiplicative structure and the action of the Bockstein are different; indeed, $H^\bullet(L(4,1),\mathbb{Z}_2) = \mathbb{Z}_2[a,b]/(a^2,b^2)$ with $|a| = 1$, $|b| = 2$, and $\beta_2 = 0$, whereas $H^\bullet(\mathbb{RP}^3,\mathbb{Z}_2) = \mathbb{Z}_2[a]/(a^4)$ with $|a| = 1$ and $\beta_2(a) = a^2$. It follows that $\widetilde{\mathcal{R}}^1_1(L(4,1),\mathbb{Z}_2) = \widetilde{\mathcal{R}}^2_1(L(4,1),\mathbb{Z}_2) = \{0\}$, yet, as noted previously, $\widetilde{\mathcal{R}}^1_1(\mathbb{RP}^3,\mathbb{Z}_2) = \widetilde{\mathcal{R}}^2_1(\mathbb{RP}^3,\mathbb{Z}_2) = \emptyset$.

EXAMPLE 7.9. For each $m, n \geq 0$, Dold constructed in **[7]** a smooth, closed manifold $P(m,n)$, defined as the quotient of $S^m \times \mathbb{CP}^n$ by the involution that acts as the antipodal map on S^m and complex conjugation of $\mathbb{CP}^n$. The Dold manifolds, which generate the unoriented cobordism ring, are orientable if and only if $m+n$ is odd. The cohomology algebra $H^\bullet(P(m,n),\mathbb{Z}_2)$ is isomorphic to $\mathbb{Z}_2[a,b]/(a^{m+1},b^{n+1})$, where $|a| = 1$ and $|b| = 2$, and the Bockstein is given by $\beta_2(a) = a^2$ and $\beta_2(b) = ab$. It follows that $\widetilde{\mathcal{R}}^q_1(P(m,n),\mathbb{Z}_2) = \mathbb{Z}_2$ if $q < m+2n$ and q is even or $q = m+2n$ and $m+n$ is even; $\widetilde{\mathcal{R}}^q_1(P(m,n),\mathbb{Z}_2) = \{0\}$ if $0 \leq q \leq m+2n$; and $\widetilde{\mathcal{R}}^q_s(P(m,n),\mathbb{Z}_2) = \emptyset$, otherwise.

7.3. A resonant characterization of orientability. We conclude this section with a result that expresses the orientability of a manifold M in terms of its top-degree resonance variety $\widetilde{\mathcal{R}}_1^m(M,\mathbb{Z}_2)$.

PROPOSITION 7.10. *A smooth, closed manifold M of dimension m is orientable if and only if $\widetilde{\mathcal{R}}_1^m(M,\mathbb{Z}_2)=\{0\}$.*

PROOF. If M is orientable, then Proposition 7.7 shows that $\widetilde{\mathcal{R}}_1^m(M,\mathbb{Z}_2)=\{0\}$. On the other hand, if M is not orientable, then the Stiefel–Whitney class $w_1=w_1(M)\in H^1(M,\mathbb{Z}_2)$ is non-zero, and

$$\begin{aligned}\delta_{w_1}(u)&=w_1u+\operatorname{Sq}^1(u)\\&=v_1u+\operatorname{Sq}^1(u)\\&=\operatorname{Sq}^1(u)+\operatorname{Sq}^1(u)\\&=0\end{aligned}\tag{55}$$

for every $u\in H^{m-1}(M,\mathbb{Z}_2)$, by the Wu formulas (54). Let $\omega=1^\vee\in H^m(M,\mathbb{Z}_2)$; since $\delta_{w_1}(\omega)=0$, this shows that $w_1\in\widetilde{\mathcal{R}}_1^m(M,\mathbb{Z}_2)$, and we are done. □

COROLLARY 7.11. *Let M be a smooth, closed, non-orientable manifold of dimension m. Suppose $H_1(M,\mathbb{Z})$ has no 2-torsion. Then $\mathcal{R}_1^m(M,\mathbb{Z}_2)=\{0\}$ whereas $\widetilde{\mathcal{R}}_1^m(M,\mathbb{Z}_2)=\mathbb{Z}_2$.*

PROOF. Proposition 7.6 shows that $\mathcal{R}_1^m(M,\mathbb{Z}_2)=\{0\}$. The proof of Proposition 7.10 shows that $\{0,w_1\}\subseteq\widetilde{\mathcal{R}}_1^m(M,\mathbb{Z}_2)$. It remain to show that this inclusion is an equality.

It follows from the proof of Lemma 7.1 and the discussion from Remark 7.2 that the map $\operatorname{Sq}^1\colon H^{m-1}(M,\mathbb{Z}_2)\to H^m(M,\mathbb{Z}_2)$ sends $w_1^\vee$ to ω and any other element to 0. Thus, if $a\in H^1(M,\mathbb{Z}_2)$ is such that $a\neq 0$ and $a\neq w_1$, then $a^\vee\neq w_1^\vee$, and so

$$\delta_a(a^\vee)=aa^\vee+\operatorname{Sq}^1(a^\vee)=\omega+0=\omega.\tag{56}$$

Since ω generates $H^m(M,\mathbb{Z}_2)=\mathbb{Z}_2$, it follows that $a\notin\widetilde{\mathcal{R}}_1^m(M,\mathbb{Z}_2)$, and the proof is complete. □

EXAMPLE 7.12. Let G be a finitely presented group which admits an epimorphism $\nu\colon G\twoheadrightarrow\mathbb{Z}_2$. As explained in [**3**], standard surgery techniques, suitably adapted to this situation, produce a smooth, closed, non-orientable, 4-dimensional manifold M with $\pi_1(M)=G$ and with $w_1(M)$ corresponding to ν under the identification $H^1(M,\mathbb{Z}_2)=\operatorname{Hom}(G,\mathbb{Z}_2)$. If, furthermore, G_{ab} has no 2-torsion (for instance, if $G=\mathbb{Z}^n$, for some $n\geq 1$), then M satisfies the hypothesis of the above corollary, and so $\mathcal{R}_1^4(M,\mathbb{Z}_2)=\{0\}$, yet $\widetilde{\mathcal{R}}_1^4(M,\mathbb{Z}_2)=\mathbb{Z}_2$.

8. Resonance varieties and Betti numbers of finite covers

In this last section we compare the Betti numbers and the resonance varieties of certain finite covers to those of the base space.

8.1. Finite, regular covers and resonance. Let X be a connected, finite-type CW-complex and let $p\colon Y \to X$ be a connected, regular cover, with group of deck transformations Γ. We assume Γ is finite, and we let $\Bbbk$ be a field of characteristic 0 or a prime not dividing the order of Γ. Given these data, a transfer argument shows that the induced homomorphism $p^*\colon H^\bullet(X,\Bbbk) \to H^\bullet(Y,\Bbbk)$ is injective, with image the subgroup $H^\bullet(Y,\Bbbk)^\Gamma$ of cohomology classes invariant under the action of Γ, see e.g. [**11**, Proposition 3G.1].

The next proposition and its corollary were proved in [**6**] in the case when $\Bbbk$ has characteristic 0. For completeness, we include a proof, following the argument given in [**33**].

PROPOSITION 8.1 ([**6**] and [**33**]). *Let Γ be a finite group and let $p\colon Y \to X$ be a regular Γ-cover. Suppose that* $\operatorname{char}(\Bbbk) \nmid |\Gamma|$, *and also* $\operatorname{char}(\Bbbk) \neq 2$ *if $H_1(X,\mathbb{Z})$ has 2-torsion. Then $p^*(\mathcal{R}^q_s(X,\Bbbk)) \subseteq \mathcal{R}^q_s(Y,\Bbbk)$, for all $i, s \geq 0$.*

PROOF. For a class $a \in H^1(Y,\Bbbk)^\Gamma = H^1(X,\Bbbk)$, the monodromy action of Γ on $H^\bullet(Y,\Bbbk)$ gives rise to an action on the chain complex $(H^\bullet(Y,\Bbbk),\delta_a)$, with the fixed subcomplex equal to $(H^\bullet(X,\Bbbk),\delta_a)$. Since Γ is finite and since $\operatorname{char}(\Bbbk) \nmid |\Gamma|$, a transfer argument shows once again that $H^\bullet(H^\bullet(Y,\Bbbk),\delta_a)^\Gamma = H^\bullet(H^\bullet(X,\Bbbk),\delta_a)$. We thus obtain an inclusion, $H^\bullet(H^\bullet(X,\Bbbk),\delta_a) \hookrightarrow H^\bullet(H^\bullet(Y,\Bbbk),\delta_a)$, and the claim follows. ∎

COROLLARY 8.2. *With notation as above, suppose the group Γ acts trivially on $H^1(Y,\Bbbk)$. Then the map $p^*\colon \mathcal{R}^1_s(X,\Bbbk) \to \mathcal{R}^1_s(Y,\Bbbk)$ is an isomorphism, for all $s \geq 0$.*

By way of contrast, the resonance varieties $\widetilde{\mathcal{R}}^q_s(Y,\mathbb{Z}_2)$ may vanish, even when the corresponding varieties $\widetilde{\mathcal{R}}^q_s(X,\mathbb{Z}_2)$ are non-zero. This phenomenon is illustrated by the next result, which is an immediate corollary to Proposition 7.10.

COROLLARY 8.3. *Let M be a smooth, closed, non-orientable manifold of dimension m, and let $\widetilde{M}$ be its orientation double cover. Then $\widetilde{\mathcal{R}}^m_1(M,\mathbb{Z}_2) \neq \{0\}$ yet $\widetilde{\mathcal{R}}^m_1(\widetilde{M},\mathbb{Z}_2) = \{0\}$.*

8.2. Mod-2 Betti numbers of 2-fold covers. We now consider the case when the order of the deck group Γ is not coprime to the characteristic of the coefficient field $\Bbbk$. We will focus on the simplest possible situation: the one where Γ is a finite cyclic group of even order (mainly, $\Gamma = \mathbb{Z}_2$) and $\Bbbk = \mathbb{Z}_2$.

Every regular $\mathbb{Z}_n$-cover $p\colon Y \to X$ is classified by a homomorphism $\alpha\colon \pi_1(X) \to \mathbb{Z}_n$, or, equivalently, by a cohomology class (called its characteristic class), $\alpha \in H^1(X,\mathbb{Z}_n)$. Furthermore, the covering space Y is connected if and only if the homomorphism α is surjective, in which case $\pi_1(Y) = \ker(\alpha)$. Note also that all 2-fold covers are regular.

We start with a simple lemma, which strengthens Lemma 3.2, part (i) from [**39**].

LEMMA 8.4. *Let $p\colon Y \to X$ be a connected $\mathbb{Z}_2$-cover, with characteristic class $\alpha \in H^1(X,\mathbb{Z}_2)$. Then p lifts to a connected, regular $\mathbb{Z}_4$-cover $\bar{p}\colon \overline{Y} \to X$ if and only if $\alpha^2 = 0$.*

PROOF. Since Y is connected, the homomorphism $\alpha\colon \pi_1(X) \to \mathbb{Z}_2$ is surjective, or, equivalently, the characteristic class α is non-zero. As noted in §5.1, the

Bockstein of α is given by $\beta_2(\alpha) = \alpha^2$. Thus, $\alpha^2 = 0$ if and only if there is a class $\bar{\alpha} \in H^1(X, \mathbb{Z}_4)$ such that α is the reduction mod-2 of $\bar{\alpha}$. Let $\bar{p}\colon \overline{Y} \to X$ be the corresponding regular $\mathbb{Z}_4$-cover. Since $\alpha \neq 0$, the homomorphism $\bar{\alpha}\colon \pi_1(X) \to \mathbb{Z}_4$ must be surjective, and so $\overline{Y}$ must be connected. □

The next result generalizes Lemma 3.2, part (ii) and Theorem 3.7 from [**39**].

PROPOSITION 8.5. *Let $p\colon Y \to X$ be a 2-fold cover, classified by a non-zero class $\alpha \in H^1(X, \mathbb{Z}_2)$. Suppose that $\alpha^2 = 0$. Then, for all $q \geq 1$,*

$$b_q(Y, \mathbb{Z}_2) = b_q(X, \mathbb{Z}_2) + \dim_{\mathbb{Z}_2} H^q(H^\bullet(X, \mathbb{Z}_2), \delta_\alpha). \tag{57}$$

In particular, $b_q(Y, \mathbb{Z}_2) \geq b_q(X, \mathbb{Z}_2)$.

PROOF. The transfer exact sequence for the 2-fold cover with characteristic class α (see [**11, 12**]) takes the form

$$\begin{aligned} \cdots \longrightarrow H^q(X, \mathbb{Z}_2) \xrightarrow{\cdot\alpha} H^q(X, \mathbb{Z}_2) \xrightarrow{p^*} H^q(Y, \mathbb{Z}_2) \\ \xrightarrow{\tau} H^q(X, \mathbb{Z}_2) \xrightarrow{\cdot\alpha} H^{q+1}(X, \mathbb{Z}_2) \longrightarrow \cdots. \end{aligned} \tag{58}$$

Since by assumption $\beta_2(\alpha) = \alpha^2$ vanishes, we have that $\delta_\alpha(u) = \alpha u$ for all $u \in H^i(X, \mathbb{Z}_2)$. Hence, the above long exact sequence splits into short exact sequences,

$$0 \longrightarrow \operatorname{im}(\delta_\alpha) \longrightarrow H^q(X, \mathbb{Z}_2) \xrightarrow{p^*} H^q(Y, \mathbb{Z}_2) \longrightarrow \ker(\delta_\alpha) \longrightarrow 0, \tag{59}$$

and the claim follows. □

REMARK 8.6. Yoshinaga [**39**] proved this result in the case when either $q = 1$ or X is the complement of a complex hyperplane arrangement. He used it to detect a $\mathbb{Z}_2$-summand in the first homology of the Milnor fiber of the icosidodecahedral arrangement.

If the 2-fold cover $Y \to X$ is classified by an element $\alpha \in H^1(X, \mathbb{Z}_2)$ with $\alpha^2 \neq 0$, the mod-2 Betti numbers of Y may actually be smaller than those of X.

EXAMPLE 8.7. For $n > 1$, let $S^n \to \mathbb{RP}^n$ be the orientation double cover, classified by the degree-1 generator $\alpha = w_1(\mathbb{RP}^n)$ of the cohomology algebra $H^\bullet(\mathbb{RP}^n, \mathbb{Z}_2) = \mathbb{Z}_2[\alpha]/(\alpha^{n+1})$. For $0 < i < n$, we have that $b_i(S^n, \mathbb{Z}_2) = 0$, yet $b_i(\mathbb{RP}^n, \mathbb{Z}_2) = 1$.

Note added in proof

After acceptance of this paper, Ishibashi, Sugawara, and Yoshinaga posted a preprint, [**13**], in which they expand on the work described in Section 8.2. Namely, suppose $Y \to X$ is a connected $\mathbb{Z}_2$-cover with characteristic class $\alpha \in H^1(X, \mathbb{Z}_2)$. Assuming $H_\bullet(X, \mathbb{Z})$ is torsion-free, it follows from [**23**, Theorem C] that

$$b_q(Y) \leq b_q(X) + \dim_{\mathbb{Z}_2} H^q(H^\bullet(X, \mathbb{Z}_2), \delta_\alpha). \tag{60}$$

When X is the (projectivized) complement of a hyperplane arrangement, an explicit formula was proposed in [**26**, Conjecture 1.9], expressing the first Betti number of the Milnor fiber of the arrangement in terms of the resonance varieties $\mathcal{R}^1_s(X, \mathbb{Z}_p)$, for $p = 2$ and 3. At the prime $p = 2$, the conjecture is equivalent to the inequality (60) holding as equality in degree $q = 1$ for the 2-fold cover $Y \to X$ corresponding to the class $\alpha \in H^1(X, \mathbb{Z}_2)$ which evaluates to 1 on each meridian. As shown in [**13**, Corollary 2.5], though, that equality holds if and only if $H_1(Y, \mathbb{Z}_2)$ has no 2-torsion. Therefore, the formula conjectured in [**26**] fails at the prime $p = 2$

precisely when $H_1(Y, \mathbb{Z}_2)$ has non-trivial 2-torsion, as is the case in the example mentioned in Remark 8.6.

References

[1] Michael Brickenstein and Alexander Dreyer, *Gröbner-free normal forms for Boolean polynomials*, J. Symbolic Comput. **48** (2013), 37–53, DOI 10.1016/j.jsc.2011.04.002. MR2980465 ↑145

[2] Michael Brickenstein, Alexander Dreyer, Gert-Martin Greuel, Markus Wedler, and Oliver Wienand, *New developments in the theory of Gröbner bases and applications to formal verification*, J. Pure Appl. Algebra **213** (2009), no. 8, 1612–1635, DOI 10.1016/j.jpaa.2008.11.043. MR2517997 ↑145

[3] Ryan Budney, *Fundamental groups of non-orientable closed four-manifolds*, MathOverflow (October 19, 2017), `https://mathoverflow.net/q/283825`. 153

[4] Daniel C. Cohen and Alexandru I. Suciu, *Characteristic varieties of arrangements*, Math. Proc. Cambridge Philos. Soc. **127** (1999), no. 1, 33–53, DOI 10.1017/S0305004199003576. MR1692519 ↑131

[5] Graham Denham, Alexandru I. Suciu, *Algebraic models and cohomology jump loci*, in preparation. 132, 136

[6] Alexandru Dimca and Ştefan Papadima, *Finite Galois covers, cohomology jump loci, formality properties, and multinets*, Ann. Sc. Norm. Super. Pisa Cl. Sci. (5) **10** (2011), no. 2, 253–268. MR2856148 ↑154

[7] Albrecht Dold, *Erzeugende der Thomschen Algebra* $\mathfrak{N}$ (German), Math. Z. **65** (1956), 25–35, DOI 10.1007/BF01473868. MR79269 ↑152

[8] Michael J. Falk, *Arrangements and cohomology*, Ann. Comb. **1** (1997), no. 2, 135–157, DOI 10.1007/BF02558471. MR1629681 ↑131

[9] Michael J. Falk, *Resonance varieties over fields of positive characteristic*, Int. Math. Res. Not. IMRN **3** (2007), Art. ID rnm009, 25, DOI 10.1093/imrn/rnm009. MR2337033 ↑132

[10] Robert Greenblatt, *Homology with local coefficients and characteristic classes*, Homology Homotopy Appl. **8** (2006), no. 2, 91–103. MR2246023 ↑146

[11] Allen Hatcher, *Algebraic topology*, Cambridge University Press, Cambridge, 2002. MR1867354 ↑144, 151, 154, 155

[12] Jean-Claude Hausmann, *Mod two homology and cohomology*, Universitext, Springer, Cham, 2014, DOI 10.1007/978-3-319-09354-3. MR3308717 ↑151, 155

[13] Suguru Ishibashi, Sakumi Sugawara, Masahiko Yoshinaga, *Betti numbers and torsions in homology groups of double coverings*, `arXiv:2209.02236`. 155

[14] Pascal Lambrechts and Don Stanley, *The rational homotopy type of configuration spaces of two points* (English, with English and French summaries), Ann. Inst. Fourier (Grenoble) **54** (2004), no. 4, 1029–1052. MR2111020 ↑143

[15] Pascal Lambrechts and Don Stanley, *Poincaré duality and commutative differential graded algebras* (English, with English and French summaries), Ann. Sci. Éc. Norm. Supér. (4) **41** (2008), no. 4, 495–509, DOI 10.24033/asens.2074. MR2489632 ↑143

[16] Anatoly Libgober and Sergey Yuzvinsky, *Cohomology of the Orlik-Solomon algebras and local systems*, Compositio Math. **121** (2000), no. 3, 337–361, DOI 10.1023/A:1001826010964. MR1761630 ↑131

[17] Samuel Lundqvist, *Boolean ideals and their varieties*, J. Pure Appl. Algebra **219** (2015), no. 10, 4521–4540, DOI 10.1016/j.jpaa.2015.02.030. MR3346504 ↑145

[18] Daniela Anca Măcinic, Ştefan Papadima, Clement Radu Popescu, and Alexandru I. Suciu, *Flat connections and resonance varieties: from rank one to higher ranks*, Trans. Amer. Math. Soc. **369** (2017), no. 2, 1309–1343, DOI 10.1090/tran/6799. MR3572275 ↑132, 138, 140

[19] William S. Massey, *On the Stiefel-Whitney classes of a manifold*, Amer. J. Math. **82** (1960), 92–102, DOI 10.2307/2372878. MR111053 ↑151

[20] Daniel Matei and Alexandru I. Suciu, *Cohomology rings and nilpotent quotients of real and complex arrangements*, Arrangements—Tokyo 1998, Adv. Stud. Pure Math., vol. 27, Kinokuniya, Tokyo, 2000, pp. 185–215, DOI 10.2969/aspm/02710185. MR1796900 ↑132

[21] Dagmar M. Meyer and Larry Smith, *Poincaré duality algebras, Macaulay's dual systems, and Steenrod operations*, Cambridge Tracts in Mathematics, vol. 167, Cambridge University Press, Cambridge, 2005, DOI 10.1017/CBO9780511542855. MR2177162 ↑142

[22] John W. Milnor and James D. Stasheff, *Characteristic classes*, Annals of Mathematics Studies, No. 76, Princeton University Press, Princeton, N. J.; University of Tokyo Press, Tokyo, 1974. MR0440554 ↑149, 151

[23] Ştefan Papadima and Alexandru I. Suciu, *The spectral sequence of an equivariant chain complex and homology with local coefficients*, Trans. Amer. Math. Soc. **362** (2010), no. 5, 2685–2721, DOI 10.1090/S0002-9947-09-05041-7. MR2584616 ↑132, 147, 155

[24] Ştefan Papadima and Alexandru I. Suciu, *Bieri-Neumann-Strebel-Renz invariants and homology jumping loci*, Proc. Lond. Math. Soc. (3) **100** (2010), no. 3, 795–834, DOI 10.1112/plms/pdp045. MR2640291 ↑132, 141, 142, 147

[25] Ştefan Papadima and Alexandru I. Suciu, *Non-abelian resonance: product and coproduct formulas*, Bridging algebra, geometry, and topology, Springer Proc. Math. Stat., vol. 96, Springer, Cham, 2014, pp. 269–280, DOI 10.1007/978-3-319-09186-0_17. MR3297121 ↑132, 141, 142

[26] Ştefan Papadima and Alexandru I. Suciu, *The Milnor fibration of a hyperplane arrangement: from modular resonance to algebraic monodromy*, Proc. Lond. Math. Soc. (3) **114** (2017), no. 6, 961–1004, DOI 10.1112/plms.12027. MR3661343 ↑155

[27] Ştefan Papadima and Alexandru I. Suciu, *The topology of compact Lie group actions through the lens of finite models*, Int. Math. Res. Not. IMRN **20** (2019), 6390–6436, DOI 10.1093/imrn/rnx294. MR4031243 ↑132, 143, 144

[28] Mikhail M. Postnikov, *The structure of the ring of intersections of three-dimensional manifolds* (Russian), Doklady Akad. Nauk SSSR (N.S.) **61** (1948), 795–797. MR0027513 ↑151

[29] Hans Samelson, *A note on the Bockstein operator*, Proc. Amer. Math. Soc. **15** (1964), 450–453, DOI 10.2307/2034522. MR164345 ↑146

[30] Larry Smith and Robert E. Stong, *Poincaré duality algebras mod two*, Adv. Math. **225** (2010), no. 4, 1929–1985, DOI 10.1016/j.aim.2010.04.013. MR2680196 ↑142

[31] Alexandru I. Suciu, *Fundamental groups of line arrangements: enumerative aspects*, Advances in algebraic geometry motivated by physics (Lowell, MA, 2000), Contemp. Math., vol. 276, Amer. Math. Soc., Providence, RI, 2001, pp. 43–79, DOI 10.1090/conm/276/04510. MR1837109 ↑132

[32] Alexandru I. Suciu, *Resonance varieties and Dwyer-Fried invariants*, Arrangements of hyperplanes—Sapporo 2009, Adv. Stud. Pure Math., vol. 62, Math. Soc. Japan, Tokyo, 2012, pp. 359–398, DOI 10.2969/aspm/06210359. MR2933803 ↑147

[33] Alexandru I. Suciu, *Hyperplane arrangements and Milnor fibrations* (English, with English and French summaries), Ann. Fac. Sci. Toulouse Math. (6) **23** (2014), no. 2, 417–481, DOI 10.5802/afst.1412. MR3205599 ↑154

[34] Alexandru I. Suciu, *Around the tangent cone theorem*, Configuration spaces, Springer INdAM Ser., vol. 14, Springer, [Cham], 2016, pp. 1–39. MR3615726 ↑132, 138, 140, 147

[35] Alexandru I. Suciu, *Poincaré duality and resonance varieties*, Proc. Roy. Soc. Edinburgh Sect. A **150** (2020), no. 6, 3001–3027, DOI 10.1017/prm.2019.55. MR4190099 ↑132, 136, 140, 141, 142, 143, 144

[36] Alexandru I. Suciu, *Formality and finiteness in rational homotopy theory*, `arXiv:2210.08310`. EMS Surv. Math. Sci., (to appear). 132

[37] Alexandru I. Suciu, *Alexander invariants and holonomy Lie algebras of commutative differential graded algebras*, in preparation. 132

[38] Alexandru I. Suciu and He Wang, *Pure virtual braids, resonance, and formality*, Math. Z. **286** (2017), no. 3-4, 1495–1524, DOI 10.1007/s00209-016-1811-x. MR3671586 ↑141, 142

[39] Masahiko Yoshinaga, *Double coverings of arrangement complements and 2-torsion in Milnor fiber homology*, Eur. J. Math. **6** (2020), no. 3, 1097–1109, DOI 10.1007/s40879-019-00387-8. MR4151730 ↑134, 154, 155

Department of Mathematics, Northeastern University, Boston, Massachusetts 02115

Email address: `a.suciu@northeastern.edu`

URL: `https://suciu.sites.northeastern.edu`

790 **Peter Crooks and Alexandru I. Suciu, Editors,** Compactifications, Configurations, and Cohomology, 2023

788 **I. S. Krasil′shchik, A. B. Sossinsky, and A. M. Verbovetsky, Editors,** The Diverse World of PDEs, 2023

787 **Hemen Dutta, Editor,** Mathematical Modelling, 2023

786 **Hemen Dutta, Editor,** Mathematical Modelling, 2023

785 **A. Leroy and S. K. Jain, Editors,** Algebra and Coding Theory, 2023

784 **Dinh-Liem Nguyen, Loc Hoang Nguyen, and Thi-Phong Nguyen, Editors,** Advances in Inverse Problems for Partial Differential Equations, 2023

783 **Alla Detinko, Michael Kapovich, Alex Kontorovich, Peter Sarnak, and Richard Schwartz, Editors,** Computational Aspects of Discrete Subgroups of Lie Groups, 2023

782 **Galina Filipuk, Alberto Lastra, and Sławomir Michalik, Editors,** Recent Trends in Formal and Analytic Solutions of Diff. Equations, 2023

781 **Emanuel Indrei, Diego Marcon, and Levon Nurbekyan, Editors,** Geometric and Functional Inequalities and Recent Topics in Nonlinear PDEs, 2023

780 **Erik Koelink, Stefan Kolb, Nicolai Reshetikhin, and Bart Vlaar, Editors,** Hypergeometry, Integrability and Lie Theory, 2022

779 **Samuele Anni, Valentijn Karemaker, and Elisa Lorenzo García, Editors,** Arithmetic, Geometry, Cryptography, and Coding Theory 2021, 2022

778 **Carlos Galindo, Alejandro Melle Hernández, Julio José Moyano-Fernández, and Wilson A. Zúñiga-Galindo, Editors,** p-Adic Analysis, Arithmetic and Singularities, 2022

777 **Bang-Yen Chen, Nicholas D. Brubaker, Takashi Sakai, Bogdan D. Suceavă, Makiko Sumi Tanaka, Hiroshi Tamaru, and Mihaela B. Vajiac, Editors,** Differential Geometry and Global Analysis, 2022

776 **Aaron Wootton, S. Allen Broughton, and Jennifer Paulhus, Editors,** Automorphisms of Riemann Surfaces, Subgroups of Mapping Class Groups and Related Topics, 2022

775 **Fernando Galaz-García, Cecilia González-Tokman, and Juan Carlos Pardo Millán, Editors,** Mexican Mathematicians in the World, 2021

774 **Randall J. Swift, Alan Krinik, Jennifer M. Switkes, and Jason H. Park, Editors,** Stochastic Processes and Functional Analysis, 2021

773 **Nicholas R. Baeth, Thiago H. Freitas, Graham J. Leuschke, and Victor H. Jorge Pérez, Editors,** Commutative Algebra, 2021

772 **Anatoly M. Vershik, Victor M. Buchstaber, and Andrey V. Malyutin, Editors,** Topology, Geometry, and Dynamics, 2021

771 **Nicolás Andruskiewitsch, Gongxiang Liu, Susan Montgomery, and Yinhuo Zhang, Editors,** Hopf Algebras, Tensor Categories and Related Topics, 2021

770 **Stéphane Ballet, Gaetan Bisson, and Irene Bouw, Editors,** Arithmetic, Geometry, Cryptography and Coding Theory, 2021

769 **Kiyoshi Igusa, Alex Martsinkovsky, and Gordana Todorov, Editors,** Representations of Algebras, Geometry and Physics, 2021

768 **Dražen Adamović, Andrej Dujella, Antun Milas, and Pavle Pandžić, Editors,** Lie Groups, Number Theory, and Vertex Algebras, 2021

767 **Moshe Jarden and Tony Shaska, Editors,** Abelian Varieties and Number Theory, 2021

766 **Paola Comparin, Eduardo Esteves, Herbert Lange, Sebastián Reyes-Carocca, and Rubí E. Rodríguez, Editors,** Geometry at the Frontier, 2021

For a complete list of titles in this series, visit the
AMS Bookstore at **www.ams.org/bookstore/conmseries/**.